DESIGNING & BUILDING MULTI-DECK Model Railroads

How to get more out of your space

Tony Koester

Kalmbach Books
21027 Crossroads Circle
Waukesha, Wisconsin 53186
www.Kalmbach.com/Books

Published in 2008
19 18 17 16 15   3 4 5 6 7

Manufactured in the United States of America

ISBN: 978-0-89024-741-9

Publisher's Cataloging-in-Publication Data

Koester, Tony.

Designing & building multi-deck model railroads : how to get more out of your space / Tony Koester

p. : ill. ; cm.

ISBN: 978-0-89024-741-9

1. Railroads--Models--Handbooks, manuals, etc. 2. Railroads--Models--Design and construction--Handbooks, manuals, etc. I. Title. II. Title: Designing and building multi-deck model railroads

TF197 .K64 2008

625.1/9

On the cover

Clockwise from far left: The author runs an eastbound freight as a westbound passes on the lower level of his HO scale Nickel Plate Road layout (photo by Robert Sobol); a scene Jack Burgess photographed showing Bagby, Calif., on the upper deck and the Merced local crossing the Santa Fe on his pioneering HO Yosemite Valley RR; and a cross-section of the author's multi-deck layout.

Contents

Jack Burgess

INTRODUCTION

Going up?

Those who want more railroad per square foot of floor space can switch to a smaller scale or – as Jack Burgess did with his superb HO depiction of the Yosemite Valley RR in August 1939 – add a second deck. Jack's craftsmanship extends from his award-winning models to his carpentry, and the railroad operates as realistically as it looks.

Model railroaders rival the great railroad barons when it comes to embracing the bigger-is-better approach to railroading. No sense filling a mere basement with a railroad in miniature when a hayloft or gymnasium will do better, we figure. But imperialistic dreams of enterprise do not magically translate to floor space. Greedy as we are, we're also a practical lot, and we understand that the family is likely to demand the lion's share of our living quarters, leaving only a modicum of space in which we can fulfill our dreams of being railroad magnates. What we therefore need is some means of multiplying the space we have, of expanding it into the vertical realm since our horizontal real estate allocation is often limited. And that's why multi-deck model railroads are increasingly a key part of the model railroader's repertoire.

Jim Hediger

The first known example of a multi-deck model railroad is *Model Railroader* senior editor Jim Hediger's pioneering Ohio Southern.

A two-for-one offer

There are quite a number of well-tested ways to gain more elbow room for your next model railroad. One of them is to switch to a smaller scale, perhaps N or even Z. You might consider a narrow-gauge railroad such as the fabled Colorado three-footers or one of the fascinating Maine two-footers, as their diminutive stature and relatively small hardware allows the use of tighter curves.

Another approach is to model an interurban line. Pick one like the Illinois Terminal or Indiana Railroad that also offered a substantial freight service and you can have the best of both worlds: frequent passenger "trains" and modelable-size freights that can negotiate curves that would give a Mack truck fits.

Most of us, however, are enamored with a particular prototype or type of railroading – perhaps hauling coal out of the Appalachians or grain out of the Great Plains. So suggestions about changing to something more manageable fall on deaf ears. Same for changing scales: We all know in our hearts that X scale is by far the best, or we rationalize our reluctance to try something new by citing our world-class roster of X scale cars, locomotives, and structures.

What we really need is a way to build twice or even three times as much railroad in our existing space. And, thanks to a creative suggestion made decades ago by the dean of track planning, the late John Armstrong, a solution is at hand: the multi-deck model railroad.

Not that John thought of more than one deck as a universal solvent, however. His own O scale Canandaigua Southern had several "levels" (it wasn't flat) but only one deck. Instead, he tossed his always thought-provoking ideas on the table and let the rest of us run with them.

The one who ran first with the multi-deck proposal was *Model Railroader* senior editor Jim Hediger, as recounted in "The ground-breaking Ohio Southern" in the September 2005 MR. Jim recalls an early 1970s meeting with John and then-MR editor Linn Westcott. Linn thought the multi-deck approach was an interesting idea and suggested Jim take a shot at it. The catch was that no one else had built one, so figuring out how to erect a second deck was up to Jim. Fortunately, Jim's a creative guy with a sound foundation in carpentry and mechanics, so he built one of the first (perhaps the very first) multi-deck model railroads, the HO scale Ohio Southern.

I have visited Jim's OS on a number of occasions over the years, and it was initially as intriguing as it was rare. Frankly, however, my interest in this approach to layout design was more professional than personal; I couldn't see myself or many other modelers adopting the multi-deck approach to layout design any time soon.

I therefore got quite a surprise during a visit to Canada in the 1990s. During an extensive layout tour, I discovered that every new or newly revised model railroad on the tour had multiple decks! One huge basement layout had at least three decks, in fact, with the upper one being well above my eye level – and I stand three inches over six feet tall (or I used to, anyway). Something was clearly in the wind.

The story of what was in the wind is what this book is all about. It's as much a fascinating story of the recent history of progressive model railroad design as it is a how-to book about the challenges created by the addition of one or more decks. I hope you find it a tale well worth your time and attention.

1
John Allen

CHAPTER ONE

The evolution of layout design

John Allen's Gorre & Daphetid HO railroad started small (see the plan on page 7, which was revised before John built the layout), but grew into a basement-size railroad (above) that incorporated the original layout. It was about as close to being a multi-deck layout as you can come without actually having a second, separate deck, and it pointed the way toward linear layout designs that embraced walkaround control.

Unless we understand how model railroad design has progressed over the decades, we can't grasp the significance of multi-deck track plans. Building a second or even third deck is a logical progression from the traditional single-deck layout, but such a simplistic viewpoint belies its potential.

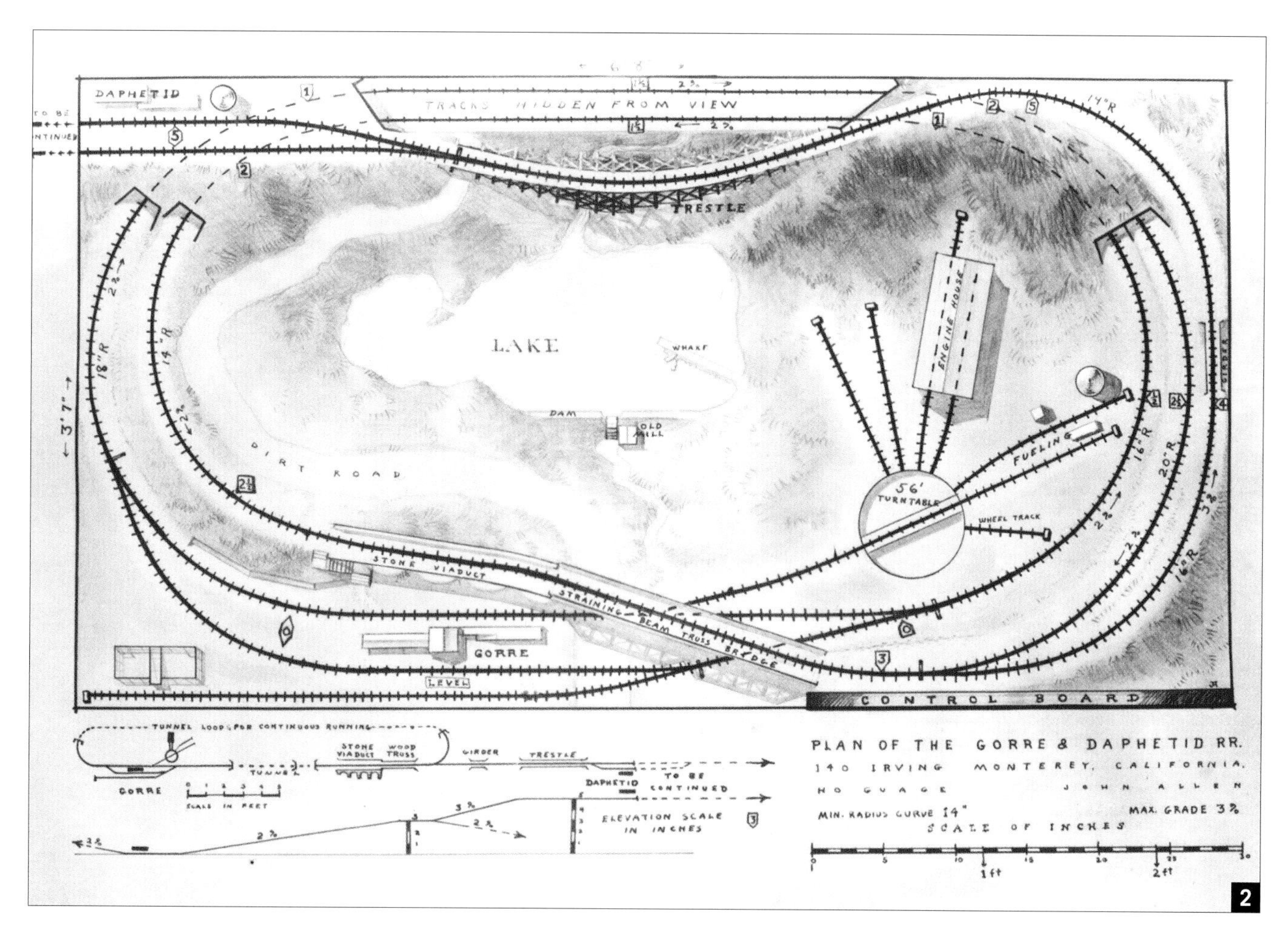

2

Indeed, multi-deck plans often fly in the face of conventional wisdom, which strives for scenic depth and grandeur, embracing instead narrow shelves that at first blush seem to defy plausible scenic treatment. So, before we discuss why one might want to build a multi-deck model railroad, let's develop a perspective on what came before. We should then more easily understand why so many new layouts are adopting the concept of multiple decks.

In the beginning

Scale model railroading has its roots in toy trains. And toy trains were sold with sectional track that allowed one to snap together an oval of track and get the trains running in no time flat, typically on the living-room floor.

The advent of 4 x 8-foot sheets of plywood allowed those who wanted to run trains year-round, instead of just at Christmas, to build somewhat permanent "layouts." Likely as not, the resulting track configuration was still the classic oval or maybe a figure 8, a convenience that continues to this day. The continuous run allowed us to enjoy seeing our trains in action for tens of minutes or maybe even hours as they tore around the confines of the good ol' 4 x 8.

One has to remember that there was little ready-to-run equipment back then, so a lot of hobby time was spent assembling locomotive and car kits that were at best highly challenging. Just getting a steam locomotive to run smoothly was a major undertaking, so seeing it run at all, let alone on a basement-size point-to-point railroad like those we commonly build today, was a major accomplishment. The very last thing on these pioneers' minds was to figure out a way to create a ten-scale-mile-long main line. The focus was, plain and simple, on building models, not on operating them realistically.

Modeling railroading

As the hobby's focus segued from toy trains to scale models of specific prototypes, it became obvious that no matter how realistic our locomotives and cars and structures might be, the arrangement of the track they ran on more closely resembled a traditional toy-train layout. Some luminary decided that a point-to-point track plan was more realistic, as it gave the engineer the sense of going from here to there, just as the full-size railroads do. Or, as Allen McClelland later put it, our railroads should look and act like a part of the North American transportation system.

One of the most important paradigm shifts in the hobby occurred when John Allen, of Monterey, Calif., built a 3'-7" by 6'-8" layout he whimsically called the Gorre & Daphetid (gory and defeated; the joke quickly got old, he later confessed). At first blush, it resembled another round-and-round layout, but the design was more sophisticated than that, **1, 2** (also see pages 42-45 of *Model Railroad Planning* 2008). In fact, the layout was good enough, with minor

Allen McClelland's original Virginian & Ohio set a course for the hobby that it follows to this day: a linear, walkaround design utilizing command control that gave train crews the feeling they were actually going somewhere instead of running around in circles. Allen operated his freelanced railroad as a link in the North American transportation network, as he described in his seminal book, *The V&O Story.*

modifications, to become an integral part of the increasingly larger second and third versions of the GD Line, but by this time John had figured out that he needed to be able to stay closer to his train, not stand like an orchestra conductor at some podium (master control panel) and view the action from afar. The aisles therefore followed the railroad as best they could as it twisted this way and that through the rugged Akinback Mountains.

Ironically, John is more famous for his spectacular scenery and weathered structures than realistic operation. However, his use of ungainly but workable Baker couplers is ample proof that he put equal emphasis on realistic operation. Why go to so much trouble to build a realistic model railroad and then not strive to operate it with the same degree of realism? That lesson is unfortunately overlooked by many of his would-be emulators who focus only on his scenic prowess.

John's ability to have his railroad operated as he desired depended on GD Line crews being able to couple and uncouple cars reliably at the desired locations. He needed to have the aisles follow closely alongside the railroad to keep reach-in distances to a minimum. That, in turn, was a giant first step toward today's highly linear layout designs, including those with more than one deck.

The third and final Gorre & Daphetid layout is about as close to a multi-deck railroad as one can achieve without actually building a multi-deck layout. John's GD Line was most certainly a multi-*level* layout, climbing from a height of 44" at Port to 61" at Cold Shoulder, with a dip down to 30" in Devil's Gulch, but it wasn't a classic multi-deck design where no effort is made to integrate the upper and lower decks. In fact, on a multi-deck layout every effort is made to visually separate the multiple decks, usually with a fascia/valance serving as a picture frame as discussed in chapter 9.

Would John embrace a multi-deck design if he were building G&D version 4.0 today? His scenic mountain vistas suggest no, as multi-decks restrict the openness of a layout's scenery. But his use of Baker couplers in days prior to the revolutionary Kadee magnetic coupler tells us that he would want to

W. Allen McClelland

have a lot of operating potential, and that increases with the length of the main line, especially if timetable and train-order operating rules are used.

It follows that John's decision would be based on the same concerns we face today: Do we have enough floor space to achieve sufficient operating potential with one deck, or will we markedly benefit by doubling or tripling the run by adding one or more additional decks?

A trend-setting model railroad

By citing one pioneer such as John Allen, I'm sure to overlook others, but this book is not intended to be a history of layout design milestones. Rather, in these pages I will single out a handful of individuals who made remarkable contributions toward the advancement of the hobby from a flanged-wheel version of table tennis to something that comes so close to the appearance and operation of prototype railroading that professional railroaders can easily relate to and enjoy it.

Among those notable pioneers is W. Allen McClelland, who in the 1960s grabbed the hobby of scale model railroading and pointed it in a new direction. Looking back at what he achieved, one has to ponder why everyone else wasn't doing what he did; most of the hobby's brighter lights are certainly following in his footsteps today.

In a nutshell, he intuitively understood that prototype railroading was really interesting, and that there were more authentic ways to model it than he had seen demonstrated elsewhere. So he built the Virginian & Ohio, **3**, and proved his point by introducing two concepts: prototype-based freelancing and modeling part of a transportation system.

Those not intimately familiar with Allen's seminal work should buy a copy of *The V&O Story,* published by Carstens Publications. The book is essentially a reprint of a series of articles that Allen wrote for *Railroad Model Craftsman* when I was the magazine's editor in the 1970s. It remains today the best reference for understanding how a milestone model railroad came to be, and hence why model railroading suddenly progressed from maestros sitting at grandiose control panels amidst their miniature and often fanciful worlds to serious amateur railroaders moving their trains from town to town and doing what

Bill Neale Collection

▲ The Railroad Prototype Modelers club in Batavia, Ill., proved that one could depict the vast prairies of the Midwest granger belt simply by modeling the right-of-way between the property fences on a narrow shelf. This photo shows one of the narrowest segments of the huge layout; little would be gained with a deeper scene.

Doug Tagsold

▲ Doug Tagsold (left) is modeling the Denver & Rio Grande Western's climb from Denver (at 38") up and around the Big Ten curves and then along the east face of the front range to and beyond the 64" summit at Moffat Tunnel on his multi-deck HO layout. Note the transition between single and double-deck at left rear. Jim Talbott works in the yard at Denver.

► Another image of Doug Tagsold's D&RGW layout shows how the lighting valance parallels the edge of the benchwork. Fluorescent lighting in the aisle and behind the valances on both decks ensure consistent, bright illumination.

looked a lot like real work, just as their full-size counterparts did.

Model work? Deal with paper forms? Learn about arcane rules and regulations? Isn't this supposed to be a hobby?

It turns out that it can be a lot of fun when you don't have to report for work at 1 a.m. on a snowy and windy night or alongside a mosquito-infested swamp in the dog days of summer, work eight to 12 hours (16 in the steam era), get a few hours sleep, and then do it all over again as the professionals did and do.

Walkaround throttles, command control, computerized Centralized Traffic Control systems, hidden staging yards, the sense of transporting cargo and people from here to there, and a freelancing concept executed so well that the model virtually became a prototype came of age in Allen's basement, and we're all much richer for it.

Between the fences

Another big leap forward in model railroad design took place in the 1970s on the Midwest Railroad Modelers HO club layout in Batavia, Ill. Some of today's most respected and progressive modelers, including Bill Darnaby, Dan Holbrook, Frank Hodina, and Bill Neale, are all alums of the Batavia club.

What Batavia showed us is that we need not be concerned with the depth of our scenery to model prototype railroading to a highly faithful and plausible degree. Instead, we can build convincing scenery simply by focusing on the relatively narrow strip of right-of-way between the fences, typically spaced 100 feet or so apart. A hundred feet in HO is about 14", so the club demonstrated that a shelf perhaps a foot to 16" wide is entirely adequate to depict the essential part of the railroad's infrastructure. High backdrops precluded seeing other operators in nearby aisles, **4**.

It wasn't a subtle demonstration, either. The first time I visited the Batavia club, my eyes popped wide open when I saw how effectively their narrow shelf layout captured the look and feel of a typical granger belt railroad. And when I participated in an operating session, the lack of great scenic depth was invisible; my eyes and brain were narrowly focused on the railroad and the train I was trying to shepherd over the railroad within the

Doug Tagsold

bounds of timetable and train-order operating rules.

And therein lies the key message of this book: Good layout design depends on the designer's scenic and operational objectives. Layouts designed primarily as scenic tours de force too often become huge dust gatherers once the scenery is largely complete, or they are quickly dismantled to make way for the Scenic Central version 2.0 and then 3.0 and 4.0. It's therefore clear that planning for realistic operation from the outset, thus allowing the "finished" railroad to be enjoyed not only sooner but also for years or even decades after it's "complete," is a very good idea indeed. And that, in turn, leads us to multi-deck layout designs.

One more point: In case you think a multi-deck design is suited only for flatlands railroading, review the accompanying photos of Doug Tagsold's Denver & Rio Grande HO layout (also see chapter 10), which depicts the climb from Denver up the Front Range of the Rockies. The Big Ten Curves section of the layout is single deck, but all other areas are built on two narrow decks. I've operated at Doug's on several occasions and can assure you that nothing has been lost from a scenic standpoint, and huge gains have been made operationally since he built and later extended the upper deck.

Multiple decks equal longer runs

If you have a relatively large basement, as John Allen and Allen McClelland did, you can build a lot of railroad in your allocated area. But both of them modeled mountain railroads, which allowed them to twist and turn their railroads' main lines up and over themselves to gain altitude and hence increased mainline length. Those who, like the Batavia club, model a flatlands railroad cannot readily resort to hill climbing to separate their railroad into several levels. And even mountain railroads like Doug's Rio Grande greatly benefit from a second deck.

Moreover, having the main line traverse a given scene more than once (as was the case with the G&D) is, to use John Armstrong's evocative terminology, less than "sincere."

What to do? Enter the multi-deck model railroad. We'll investigate whether you actually need to embrace the added complexity and cost of a multi-deck layout in chapter 2.

1
Bill Darnaby

CHAPTER TWO

Do you really need more than one deck?

Bill Darnaby's trend-setting Cleveland, Indianapolis, Cincinnati & St. Louis (the Maumee Route) was one of the first home layouts to employ a multi-deck design to increase the mainline run for enhanced timetable and train-order operations. Deck heights range from 38" at East Yard in LaFontaine, Ohio (the depot is visible at lower right), to 66" at Dacron, Ohio, as the main line gently climbs eastbound. Adding a second deck allowed Bill to double the mainline run to ten scale miles (about 600 feet) in his 35'-2" x 45'-1" L-shaped basement.

The most telling comment about why multi-deck model railroads have become so popular was a statement Bill Darnaby made when he was coaching me to abandon my phobia against multi-deck designs as I struggled with the plans for my new HO railroad.

My goal was (and is) to depict the eastern half of the former Nickel Plate Road's St. Louis Division late in the steam era. That railroad ran by timetable and train-order rules and moved freight behind fleet-of-foot Berkshires, which could gobble up the miles – actual or scale – in no time flat. Obviously, I needed a long mainline run. Bill faced the same concerns when he designed his Maumee Route, **1**, and he managed to fit 600 feet – ten scale miles – of HO railroad into his L-shaped basement. He urged that I follow his example. "But I don't really care for multi-deck layouts," I countered. "Neither do I," he replied, and then added, "but I like what they let me do."

Seeing the entire "movie"

My initial track plan for the new layout started out in a staging yard just east of Frankfort, Ind., one of the Nickel Plate Road's two major hubs. Trains would enter Frankfort and be reclassified there before continuing westward over the Third Subdivision of the St. Louis Division, **2**. Then they would run over the Indiana half of the Third Sub to the Illinois state line and – poof! – disappear into another staging yard. "I wouldn't do that," said Bill.

Bill described a scenario that begins as engine crews "get on" their engines on the ready track by the coal dock, **3**, trundle down to the eastbound yard office, **4**, to get their Clearance Form A and any train orders or messages, chuff down to the west end of the westbound yard, back onto their train, and wait for a signal to leave the yard via the interlocking at WY Tower, **5**. Out on the line, they continually refer to their copy of the employee timetable and train orders, if any, to see whether they can make it to the next town, and then the next town after that, without delaying a superior eastbound train.

Then, with my proposed track plan, just about the time the train crew is getting comfortable with the task at hand, the train mysteriously vanishes off the face of the Earth into a hidden staging yard just beyond the state line. Wouldn't it be better, Bill asked, for them to complete the run to the next division-point yard at Charleston, Ill., then yard their train, put their engine on the inspection pit, and give themselves a high-five to celebrate a successful run over the entire division?

Needless to say, I was so excited about the prospect of this all happening in my basement that any resistance to the idea of a multi-deck layout was permanently relegated to a mental file folder marked "so 20th century." I instantly became a huge fan of multi-deck model railroads, as with few exceptions they almost double or even triple your fun in the same footprint.

Reverse engineering

Bill was right: To meet my operational goals, I really did need to have almost twice as much mainline run as my original single-deck track plan allowed for. In fact, as I look back on our discussion now, when the railroad is fully operational and has met my operating goals, it would have been downright silly of me to forfeit a 100-percent gain in mainline run just because I had some old-school hang-ups about adding a second deck.

In retrospect, I can see where I made a basic mistake in my initial track-planning endeavors: I focused on fitting the railroad into the room instead of first determining what the end result had to be. I had operated on Bill's Maumee, and I knew without a doubt that his multi-deck approach worked extremely well. I also knew that his long (ten-scale-mile) mainline run was a key part of the success of his design and subsequent operations. So why did I continue to insist on a single-deck plan that generated at best a 250-foot mainline run?

I suspect that the model railroad that previously occupied the same basement, the freelanced Allegheny Midland, had a lot to do with it. The multi-level (but not multi-deck!) AM was set in the Appalachians, and the mountain ridge that ran along the spine of the central peninsula was scenically attractive. It was high enough to block one's view of action on the other side of the ridge, yet low enough for operators to keep in touch with one another in a social sense.

All of which had nothing to do with the design requirements for the Nickel Plate layout. In fact, since the whole idea was for train crews to make decisions about whether it was safe to proceed by consulting the timetable and train orders (if any), the very last thing I wanted was for someone to be able to look over the peninsula (or under it, thus utilizing "sneaker orders") to see whether an opposing train was coming. That is, grand scenic vistas were a potential handicap, not a goal.

Moreover, in a timetable and train-order operating environment, the distances between passing tracks and the number of such refuges become key design elements. You want train crews to have numerous passing tracks where they can duck out of harm's way for superior trains, but you don't want them so close together that they can just run on "smoke orders" – visually determining whether they can make it safely to the next siding.

Instead, you want them to develop a feel for times and distances and train performance, and make movement decisions based on the book of rules and the superiority or lack thereof of their trains, as I described in my book, *Realistic Model Railroad Operation* (Kalmbach Books).

Had I been modeling a railroad that ran fewer trains or operated at a much slower pace, as many shortline and narrow-gauge railroads did, then fewer passing tracks and/or a shorter main line might have been acceptable. But I wasn't, so they weren't.

The bottom line is that one has to do some reverse engineering – working backward from the desired end result – to create the proper environment to ensure that result is eventually achieved. And that may lead to a multi-deck track plan.

Veteran modeler John Swanson, who models a granger railroad set in the 1920s, agrees: "It would have been far better if I had left out several towns, especially the longer ones, and had instead increased the length of run between towns. I allowed a minimum run of 15 feet between towns but should have increased that to at least 25 feet. Having run on several layouts

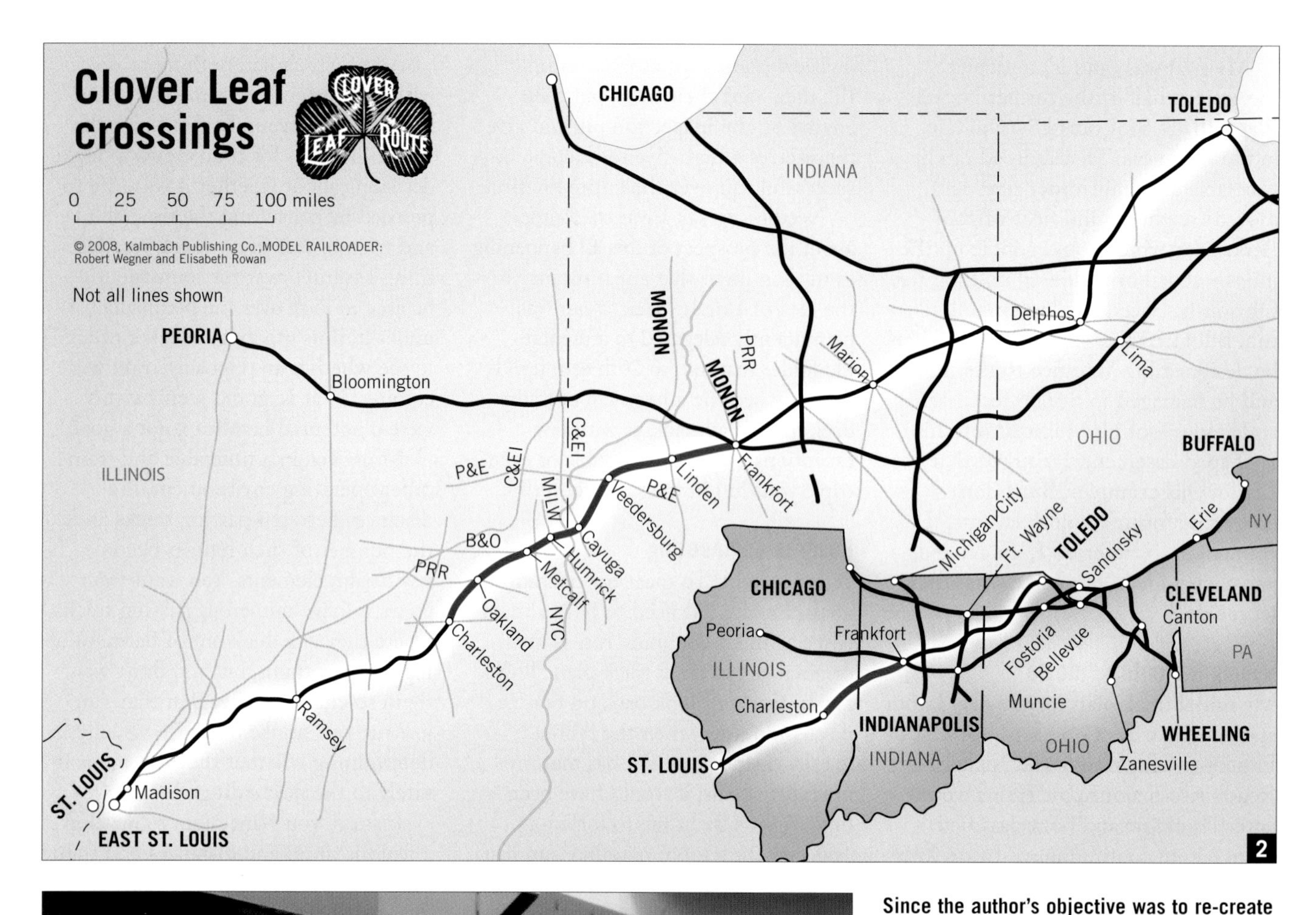

2

3

Since the author's objective was to re-create the operations of the Third Subdivision of the Nickel Plate Road's St. Louis Division (see map above) in the steam-diesel transition era, it followed that modeling both division-point yards would enhance the train crews' over-the-road experience. This meant "getting on" their engines by the coal dock at Frankfort, Ind. (3), stopping by the eastbound yard office (4) for orders and/or messages, getting on their trains, and then passing WY Tower (5) as they headed for Charleston, Ill. This is an eastbound entering Frankfurt Yard from the west.

4

5

Givens and druthers

Here's a sampling of must-haves and nice-to-haves you might consider:

Givens:

- Walkaround, point-to-point track plan
- Long mainline run
- Staging yard at both ends
- Numerous interchanges
- Several bridges
- Five towns
- Classification yard
- Realistic operation by timetable and train order
- Digital Command Control radio-controlled throttles
- Steam-diesel mix
- Center-of-the-room stairs

Druthers:

- Crew lounge with bathroom
- Wide aisles
- Two yards and engine terminals
- Dispatcher and agent-operator
- Carpeting
- Air-conditioning
- Continuous climb between decks

where the multi-deck concept was used to increase the length of the run instead of cramming in more towns, I have found the longer run to be a much better choice. Pulling into one town just as your caboose leaves the previous town doesn't give you the feel of running in open country."

Where to begin

A good place to start layout planning is by writing down a statement, preferably in list format, that clearly documents your goals. John Armstrong called them "givens and druthers" – the elements you absolutely have to have, and those that would be nice to have (see above).

In my case, for example, among my givens were the desire to model a portion of the NKP at the end of the steam era, **6**; operate using timetable and train-order rules with enough mainline run to allow road crews to face many of the same operational challenges, but none of the dangers, of their professional peers; re-create as many railroad jobs as possible except those that were purely paper-shuffling administrative tasks, **7**; Digital Command Control with radio throttles and sound decoders in all locomotives, **8**; sufficient staging to operate an entire 24-hour day, **9**; and bright, even lighting (see chapter 6).

Druthers included freight trains up to 30 cars long, **10**; a comfortable operating environment with a crew lounge; four-hour operating sessions; operators familiar with timetable operations; manageable crew size (between 12 and 16 operators); convincing Midwestern scenery, including cash crops, **11**; and monthly operating sessions.

Once your criteria are on paper where they can be referred to, then you should reflect on whether you have the resources to get there from here. Does the prototype you're modeling (or the prototype(s) you're using as a benchmark for freelancing) support the type of railroading you most enjoy, **12**? If your favorite railroad had a double-track main, **13**, but you're intrigued by timetable and train-order operations over a busy single-track main line, you have a tough decision to make.

Do you have enough space to model a sufficiently long portion of that railroad? Bear in mind that you could model nothing but one or more classification yards that interconnect or are fed by transfer runs and trains arriving and departing from one or more adjacent staging yards, **14**. Or you could eschew yards entirely, thereby saving considerable space and cost, by modeling one or a few towns along the line and running trains through them that come out of and go into nearby passive staging or active fiddle yards. This is somewhat like the exhibition layouts in the United Kingdom, which often feature single towns fed by off-stage fiddle yards, **15**.

Are there enough model railroaders interested in operating your favorite brand of railroad available to help you re-create the operating scenario you envision? The prospects may be better than you think, as modelers routinely drive two or more hours each way to attend well-conceived and well-executed operating sessions.

Conversely, if you operate alone or with one or two friends, are you planning more railroad than you can comfortably build, maintain, and operate?

Can you find and afford the type and quantity of locomotives, cabooses, and passenger cars needed to bring a favorite railroad or flavor of railroading to life? If major kitbashing or even scratchbuilding projects lie ahead of you, you may want to scale down the size and complexity of your layout to balance the time requirements.

Are the models up to the job – that is, can a 2-8-0 Camelback, **16**, or General Electric 70-tonner pull enough loaded hopper cars or brass passenger cars up the railroad's ruling grade – which could be a helix that climbs between decks, as we'll discuss in chapter 4? Can they be fitted with Digital Command Control sound decoders and speakers, which are increasingly viewed as the norm?

Modeling jobs

The concept of linking "modeling" to "jobs" takes a bit of getting used to. The idea is that there is more to railroading than neat scenery and equipment. Learning how railroads earned their keep is a fascinating study

6

Darrell Finney; author's collection

Frankfort, Ind., one of two hubs on the Nickel Plate Road, saw a mix of steam and diesel power in the early to mid-1950s. This mid-'50s portrait by NKP dispatcher Darrell Finney of Berkshires facing down invading EMD GP9s depicts the era the author is modeling.

of industrial archeology, just as an understanding of basic geology helps us plan and build better scenery, as I discussed in *Planning Scenery for Your Model Railroad* (Kalmbach Books).

More to the point, the whole concept behind my new NKP-based layout is focused on re-creating the working practices of the steam-to-diesel transition era of the 1950s. Road crews can be on duty up to 16 continuous hours. Over-the-road trains can typically stop to work at no more than two places without incurring extra wages for road crews. The nuances of timetable and train-order operation require a thorough working knowledge of the book of rules. Agent-operators in each town were the customers' link with the railroad and train crews' link with the dispatcher, **17**. Telephones and telegraph, not radios, were the primary means of communication.

The main reason for dismantling the Allegheny Midland after 25 years of construction and operation was to re-create (and thereby honor) part of a railroad, including its operating practices, that has fascinated me

7

Wallace W. Abbey

Modelers have discovered that one of the more interesting jobs to "model" is that of an agent-operator, who works closely with the dispatcher and yardmasters to ensure that train crews know what they are expected to do.

▲ **The author uses Digital Command Control with handheld radio throttles from NCE. Locomotives are equipped with sound decoders that allow crews to send or acknowledge whistle signals that, for example, a section is following.**

▼ **The author's NKP Third Sub has enough staging tracks at both ends to allow an entire 24-hour-day's schedule to be accommodated, although each session simulates only half a day – four actual hours using a 3:1 fast clock.**

since I was nine years old. I couldn't accomplish that by modeling only the scenery, structures, locomotives, and rolling stock, no matter how perfectly they were crafted. I also had to find out about and then emulate the way the railroad went about its daily business.

I believe model railroading is the closest thing to a time machine that's been invented. Done right, our model railroads can replicate almost all of the actions of their full-size counterparts from the look and sound of a locomotive to the purposeful movement of freight and passenger cars.

Modelers today therefore increasingly strive not only to get the nuts and bolts of the rolling stock and structures correct, but also to emulate the way the railroad conducted its business. That is, we now "model jobs," to use Layout Design Special Interest Group (LDSIG) founder Doug Gurin's apt term.

On the NKP, for example, I need to fill the following jobs before an operating session can commence:

• Dispatcher, who observes how trains are progressing against the published schedule and issues train orders to amend the status quo as required.

▲ The train-length goal for the Third Sub was 30 cars, which translated to 18-foot-long passing and staging tracks. Since the railroad climbs westbound as it spirals around the basement, free-rolling trucks with metal wheels are mandatory to allow steam locomotives to handle such long trains.

▼ Crops such as soybeans (shown here), corn, and wheat are the bread-and-butter of granger railroads such as the Nickel Plate Road, so they need to be represented. The bean field at Linden, Ind., was modeled using Woodland Scenics foliage clumps; the rows were spaced to match Athearn's John Deere tractor.

Whether you model a railroad such as the Delaware & Hudson because of its classy paint scheme or locomotives (above) or the Pennsylvania Railroad and its successors because of its traffic density and scenery – as Ken McCorry did in HO scale (below) – carefully defining your givens and druthers and the resources needed to model it are critical.

Brad Bower

• Two agent-operators, who copy train orders issued by the dispatcher and give them to train crews and determine how best to meet the needs of local industries – such as the number of empties needed for loading today.

• An eastbound and a westbound yardmaster at Frankfort, Ind., plus a general yardmaster who coordinates the workflow between the two as they build or break up approximately 20 trains per 12-hour "trick" for the four divisions that radiate out of Frankfort.

• A west-end crew at Frankfort to work industries and the adjoining Peoria Division staging yard and truncated main line.

• An industrial engine crew at Frankfort, which switches industries and interchange tracks at the east end of town and handles trains into and out of both the Toledo and Sandusky Division staging yards.

• An engine hostler at both Frankfort, Ind., and Charleston, Ill., who moves engines to and from the inspection pit and turns them or stores them in the roundhouse between runs. He also assists the industrial engine crew as needed.

• A yardmaster at Charleston, the other division-point yard.

• Enough two-person Third Sub road crews to operate the one passenger and approximately half-dozen freight trains per 12-hour trick, including the local, plus any additional sections or extras.

Sounds like a lot of work, you say? That's the idea! And it explains why I've become such a fan of multi-deck model railroads.

I'm fortunate that a large number of modelers who are willing to dig into the arcane art of timetable and train-order operation live within a two-hour drive of my home. Had the available pool of experienced operators been smaller, I could have reverted to single-person road crews, combined some of the hostling and staging jobs, and otherwise simplified operations. In fact, when attendance is light, we employ one-person road crews and eliminate the general yardmaster job at Frankfort.

There is a rapidly increasing number of modelers who have discovered the

Jim Senese's Kansas City Terminal Ry., featured in *Model Railroad Planning* 1999, features transfer runs between four major classification yards, each operated by a different railroad, as well as several major industries bracketing the Missouri-Kansas state line.

John H. Wright

HO fine-scale modeler John H. Wright has built several superb exhibition layouts. As described in *Model Railroad Planning* 2008, he is now building a new layout divided into two sections on either side of a central staging and control area.

David P. Morgan Library collection

Nothing says "eastern anthracite coalfields" quite as eloquently as a Camelback steam locomotive, such as Reading 2-8-0 No. 1613, with its mid-boiler cab occasioned by the wide firebox for slow-burning hard coal. But are affordable, good-pulling models of favorite prototypes of such locomotives available?

fun of realistic operation, usually by exposure to another operating layout in their area. Within 25 miles of my home, the number of layouts designed for realistic operation is about to triple. Modelers have demonstrated that they will drive scores of miles to operate on model railroads that reflect the realism and operating practices of the full-size railroads, another example of the "if you build it, they will come" syndrome. Clearly, a multi-deck model railroad offers more of everything and hence more opportunities for more modelers to participate.

Look – two of everything!

No matter how you look at it – more scenery and structures or more main line – if you decide that adding one or more decks to your railroad is advantageous, you will sooner or later have to come to grips with something you innately know but still fail to understand properly at this juncture: You have to build the railroad twice – once on each deck.

I clearly recall feeling highly satisfied, even elated, about completing the infrastructure of the lower deck of my current model railroad. But as I stepped back to admire my handiwork, smug in the knowledge that I had solved several pressing engineering and aesthetic concerns, it gradually sank in that I now had to do the whole thing all over again on the second deck.

Moreover, the second deck would be far more challenging for reasons that will become more apparent in chapters 6 through 8. Consider, for example, the fact that all wiring and switch motors for the upper deck hang down into the "sky" area of the lower deck, as do lower-deck lighting fixtures.

This can throw off your instincts by a considerable margin. If you've built a model railroad of a similar footprint before, you have some sense of what it will take to build another railroad of the same size. But if you're building a multi-deck layout, you have to multiply everything – time, money, lumber, wiring, track, scenery, backdrop – by two or maybe even three.

Although most lower-deck construction is typically quite simple, the engineering challenges presented by a second or third deck can be daunting. Consider the large open span at Cayuga, Ind., on my HO layout (see photo 5 in chapter 8) or the two-deck swinging gate that provides access to the water softener and heater (photo 4 in chapter 3).

But model railroading is all about solving challenges, whether you're trying to find photos or critical dimensions for a model, wire a signaling system, or build cantilevered upper-deck benchwork. The most important consideration is whether the ends justify the means. If they appear to, then press on, as the problems can be solved. If you're uncertain about the value proposition, as most readers of this book are likely to be, it's time to think long and hard about your available resources, including whether experienced friends will be available when the going gets tough.

Proof of the pudding

As you look through this book, you'll find numerous photos of my own multi-deck model railroad. Even without a lot of pretty scenery to look at yet (assuming you call corn, soybean, and wheat fields "scenery"), that should give you a large clue as to which side of the argument – to multi-deck or not to multi-deck – that I support.

I won't kid you into believing it has been a breeze since Bill Darnaby dramatically changed my attitude toward multi-deck layouts, but I can assure you that I made the right choice. In fact, I'll go so far as to say that the railroad would have been a failure when measured against my operating objectives had I built a single-deck design.

Even after embracing the idea of twice as much railroad in a given footprint, I remained very apprehensive about whether I had enough mainline run to achieve the desired level of timetable and train-order operation on a high-speed, single-track railroad. The plan (shown on pages 26 and 27) crafted by expert track-planner and fellow Nickel Plate enthusiast Frank Hodina, who earned a degree in railroad engineering, was able to achieve "only" eight scale miles (about 500 actual feet) of main line. Bill Darnaby's Maumee has a ten-scale-mile (600-foot) main line.

17

Judy Koester

The agent-operator at Charleston, Ill., on the author's layout relays train passing (OS) reports to the dispatcher, copies train orders and messages, and ensures that local shippers have the cars they need for loading. A second operator at Frankfort, Ind., handles similar duties in the other main aisle.

The question was very simple: Would 80 percent of the Maumee's mainline run provide enough railroad to emulate the highly successful operating scheme Bill had achieved? The answer (as Bill assured me it would be from the outset) is a resounding Yes!, but it took me the better part of seven years (I work very slowly at times) to find out for sure.

At this writing, the railroad has been operating regularly for the past year, and by all reports from the crews we have achieved something unique: a time machine that lets us experience what professional railroaders did and what local railfans (including me) saw on the Nickel Plate Road's St. Louis Division more than half a century ago.

I won't recommend that you blindly follow Bill's and my example by building a big, multi-deck model railroad, or even a small one that fits along one wall of a spare room. But I will strongly recommend that you put any apprehensions about the validity of the multi-deck approach to layout design aside and instead investigate your operational and scenic objectives.

Then decide whether you can achieve them with a single-deck track plan. If not, embrace the multi-deck approach, take a deep breath, and follow along as we discuss what you need to consider and what you need to do to get there from here.

Brad Bower

CHAPTER THREE

Design considerations

Four decks of Ken McCorry's ambitious HO layout – two scenicked, two for staging – are visible in this photo. He cautions track planners to think about maintaining their layouts as they get older, especially deep scenes that require reaching into tight places.

No matter how earnestly we pursue a quest, we will seldom be able to accomplish our goals on our own. This teamwork can assume many forms: a round-robin work group to speed up progress, a few trusted friends who have skills we lack and who buy into what we're hoping to accomplish, and a sufficient pool of operators to help us run the railroads we singly or collectively create. Even the height and physical mobility of those who will help us through any phase of model railroad design, construction, and operation may directly affect our track-planning endeavors. A layout built to accommodate a wheelchair or the grandkids may have different height requirements than the norm.

Finding a plan

I don't recommend blindly copying a published track plan, as I think you can do better by borrowing bits and pieces from actual prototype locations and connecting them together like pearls on a string. I dubbed these bits Layout Design Elements (LDEs) and discussed them at length in my book, *Realistic Model Railroad Building Blocks* (Kalmbach Books). The basic idea is that by copying in simplified form prototype track arrangements for towns, yards, engine terminals, junctions, industries, and so on, you can be pretty sure that they will look good and operate well, even if you still don't fully understand their real-world function.

Although numerous examples of multi-deck track plans have been published (some of which appear again in this book), keep in mind that most single-deck track plans can be converted into multi-deck plans by adding a helix at one end, or by having the main line climb almost continuously to reach the upper deck. The second deck gives you the chance to add 100 percent more of the type of railroading that initially attracted you to the single-deck layout's plan.

Planning ahead

We'll investigate layout height in chapter 5 and lighting in chapter 6, but at this early point in the design process, it's worth reflecting on the inevitable: aging. Tomorrow gets here faster than we expect. Unless we plan for what's coming, we're likely to make some design decisions we'll sooner or later come to regret.

And no matter how young we are, circumstances may make it difficult for us to nod, duck, or crawl under lower benchwork. Many modelers position a caster-equipped chair near duckunders to convert a painful bend into a short ride.

Ken McCorry put it this way: "There were places on my multi-deck HO layout, **1**, that I could no longer reach, probably due to my becoming 'thicker' as I get older, so I had to build removable scenery panels. If there's a good chance that the railroad you're planning or building today will have to accommodate an older you, consider the likelihood of increased limitations on reaching in, bending over, and so on. If you build higher than armpit height, consider a raised floor in that area."

Many design decisions are not specifically related to a multi-deck railroad but simply reflect good practices. You've been warned time and

A wine cellar under the front porch proved just wide enough to house local industries on the top deck and the coal branch below it on Perry Squier's circa-1923 Pittsburg, Shawmut & Northern. This area is reached by the duckunder at the left, a major gain for a small inconvenience.

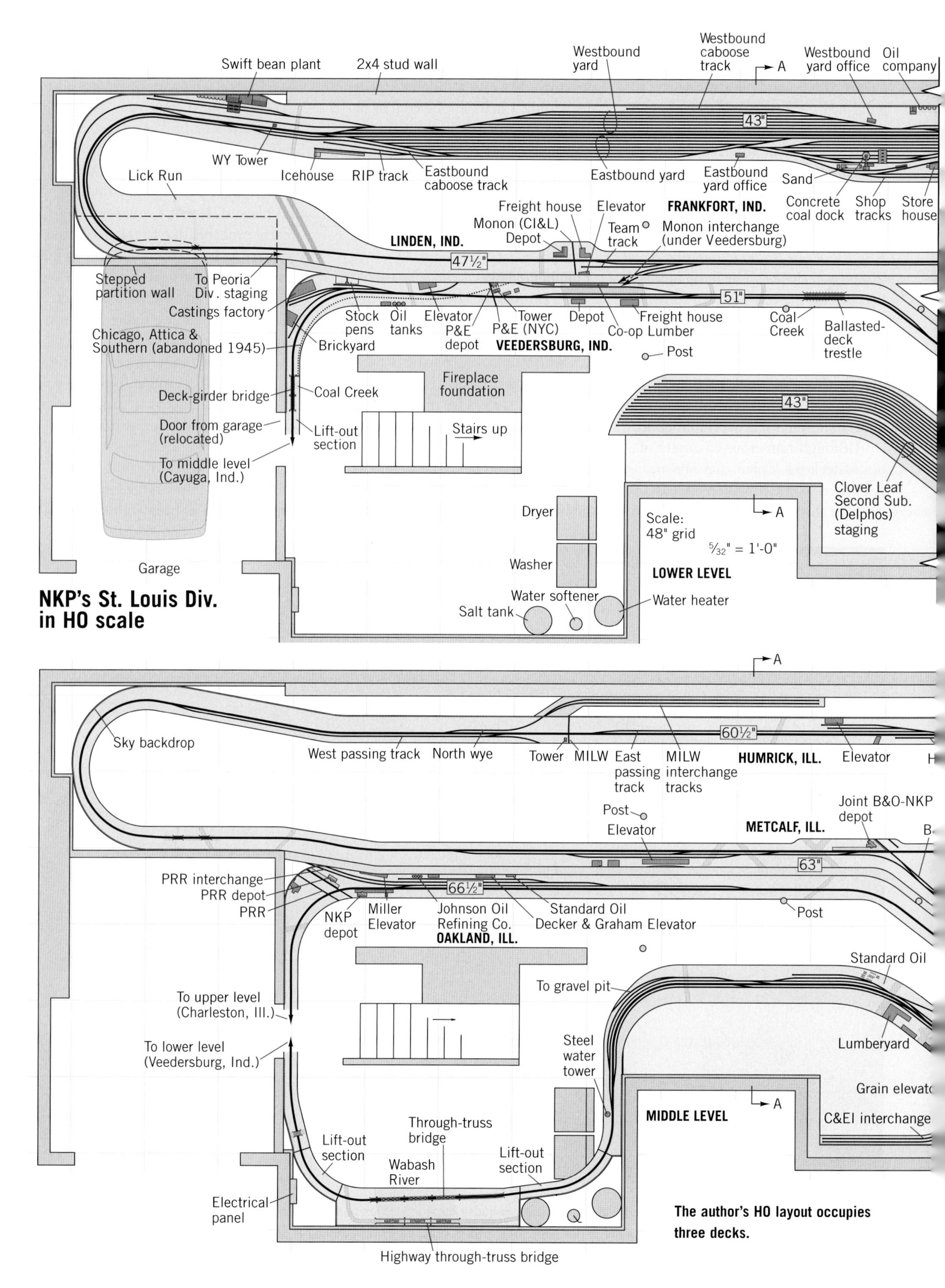

The author's HO layout occupies three decks.

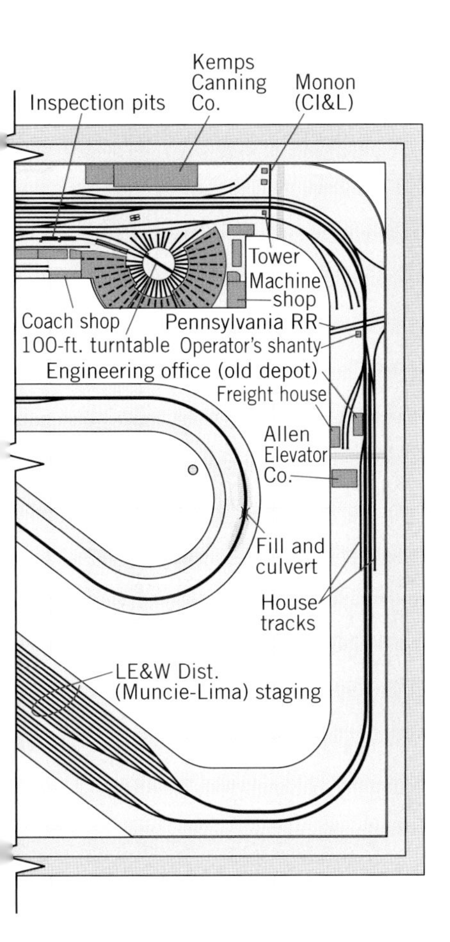

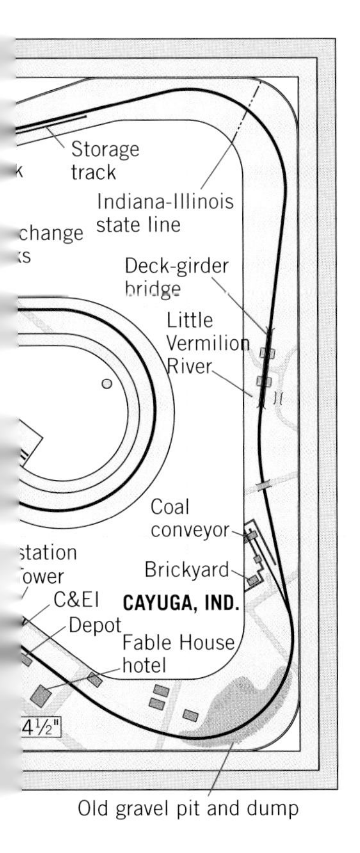

There are two duckunders on the author's NKP – one with 48" of clearance to reach the Charleston yardmaster's alcove (top) and another 42" high to enter the dispatcher's office inside the peninsula's "blob" (above). Neither has to be negotiated by road crews.

3

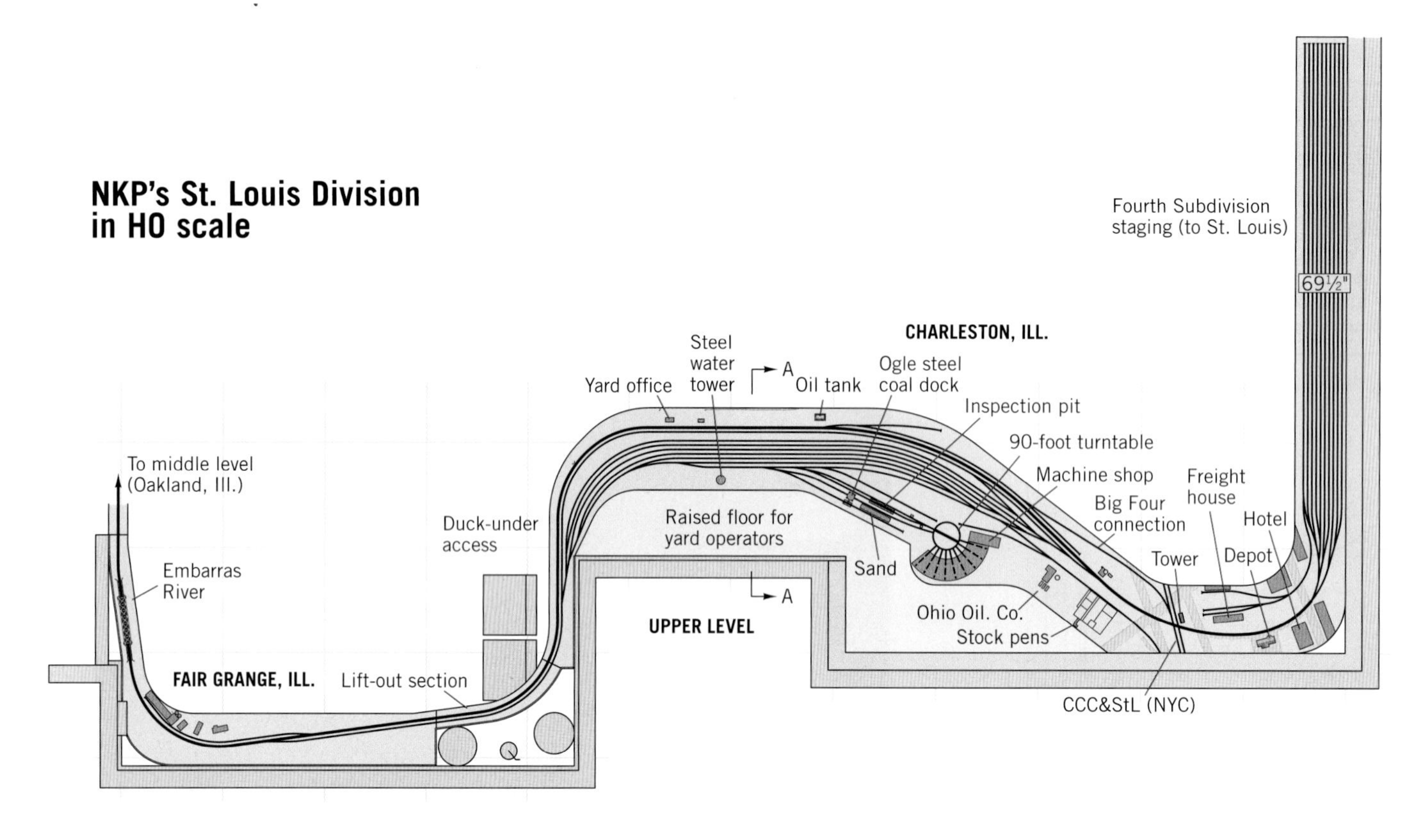

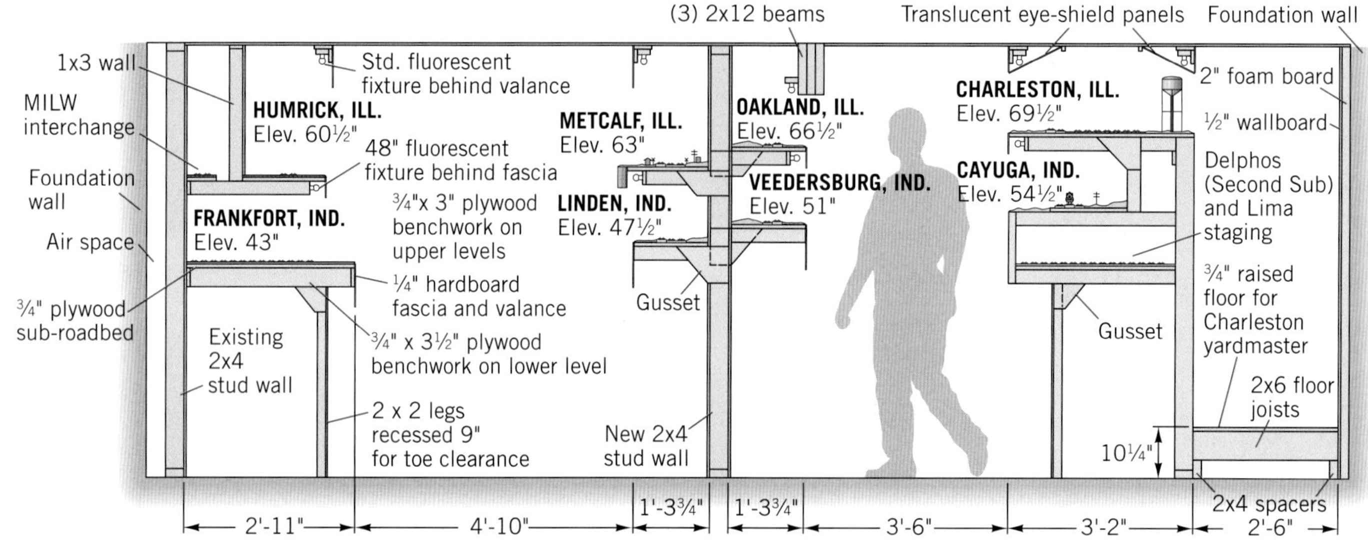

CROSS-SECTION A-A

time again about the undesirability of duckunders. This tends to be a bigger problem with a multi-deck layout, as the bottom deck is usually lower than the base height of a single-deck layout to avoid having the upper deck above eye level, **2**, **3**.

A good way to avoid duckunders is to build the railroad in a room where the entrance is via stairs in the center. My own railroad (see track plan on page 26) reflects this design, something I watched out for when our home was being built back in 1973. I wasn't as proactive when it came to locating the water heater and softener, however, so I had to build a swing-out gate to provide access to them, **4**.

Bill Sornsin had a special building constructed for his multi-deck HO Great Northern layout, which includes a garage, band room, and basketball court on the ground-level roof. Where a duckunder seemed inevitable, Bill located a set of stairs in a well that allows easy transit under the benchwork, **5**. Neal Schorr took a similar approach with his single-deck O scale Pennsylvania RR.

There is another entrance to our basement from the attached garage, but during operating sessions it is blocked by two removable lift-outs, **6**. These sections simply drop into place and are connected to the main line with short sections of removable flextrack on both ends.

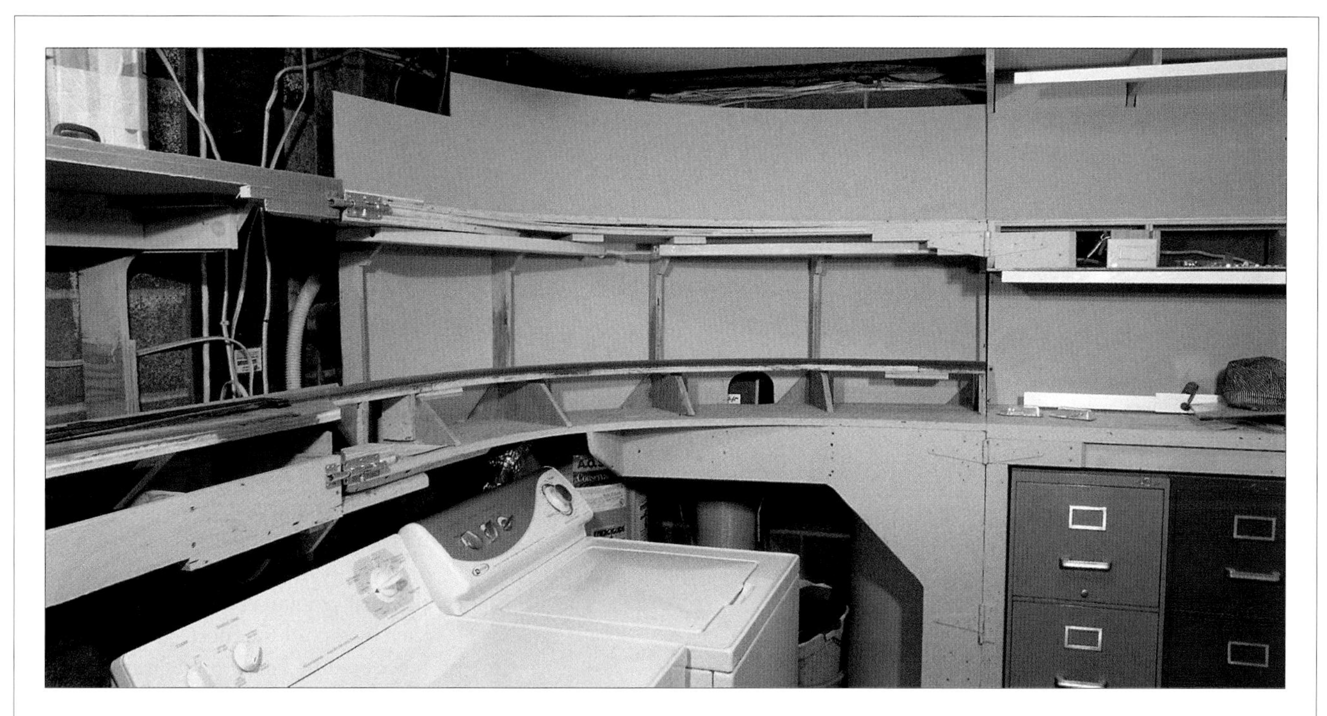

To provide access to the water heater and softener, the author constructed a bi-level gate, described in detail in April 2005 MR. It's supported by the hinged, L-shaped plywood base, and the curved hardboard backdrop provides stiffening. Five C-shaped plywood brackets support the upper deck and under-cabinet lighting. Short, removable sections of flextrack span the joints.

4

An operating alcove

A key attribute of Frank Hodina's plan for my railroad was to move Charleston yard (as well as the town of Cayuga below it, and the east-end staging yard below Cayuga) out from the basement wall (page 28). This provides an alcove for the Charleston yardmaster to work in without being hassled by crews in the main aisle. This is the high end of the railroad, so I built an elevated floor to make it easier to work in this yard, as documented in chapter 5.

The only problem with this innovative design is that the yardmaster and Charleston engine hostler have to duck under the middle (and upper) deck, **5**. The lowest point of the duckunder is 48" above the carpeted floor, so it's not a crawl-under but not

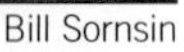
Bill Sornsin

Neal Schorr

When Bill Sornsin designed a special building for his multi-deck Great Northern HO layout, he included a stair pit to convert a duckunder into a nod-under (top). Neal Schorr used the same approach for his 3-rail O scale Pennsylvania RR layout and even included PRR-style fencing on either side of the stairs (above).

5

a more-desirable nod-under either. I think it's a good trade-off.

The dispatcher also has to duck-walk under the lower deck as he or she enters the dispatcher's office in the "blob" at the end of the main peninsula, **3**. If this becomes an issue, I'll move the office to an open area under the basement stairs now used for model storage. Chapter 5 discusses ways to navigate low duckunders.

Efficient track plans

Modelers who have studied track planning in depth have concluded that the most efficient design for a model railroad is an around-the-walls plan with a central peninsula. Such plans waste the least amount of space on aisles and thereby accord the most space to the model railroad itself for a given floor area. It follows that a multi-deck layout built to this design will be even more efficient, as you get twice as much railroad for the same amount of aisle space.

Mushroom-style layout pioneer Joe Fugate has devoted a section of his Web site (siskiyou-railfan.net) to a discussion of layout-design parameters, including articles on whether your layout will work before you build it, layout design analysis, statistics on operating potential, estimating building time and cost, and so on. We'll discuss his accomplishments a bit more in chapter 4.

L-shaped layouts

My experience operating on Bill Darnaby's Maumee Route has convinced me that the ideal shape for a railroad featuring timetable and train-order operation is an L or even a U. The bends at the corners of the L or U and at the ends of the central peninsula, if any, provide long runs between towns and virtually ensure that train crews can't move ahead by resorting to smoke orders (looking down the aisle to see whether the way is clear) instead of doing due diligence.

The whole point of timetable and train-order operation on a model railroad is to enjoy solving the time-distance problems presented by the schedule, sort of like chess against a clock. Track plans that keep you from even subconsciously looking ahead are therefore a plus.

Keep this in mind the next time your real estate agent or architect shows you a wide-open floor plan for the area you have earmarked for the new railroad. For basement or attic railroads, be sure the stairs are located somewhere away from any wall.

Are engines up to the job?

I mentioned this briefly in an earlier chapter, but now's the time to get out some pieces of flextrack and create a test grade as steep as the ruling grade on your railroad. This is likely to be the helix between levels if you plan to have one. If that grade is on a curve, which is also likely, add such a curve to the test section. It needs to be at least a train length and a half long so the full weight and "drag" of the train is on the grade.

Then assemble a sample train of the planned length (does it look long enough to emulate the prototype?) and see whether the locomotives you plan to use can actually handle that train on that curving gradient.

You're not quite done yet. Try running that test train down the grade. Do the cars push the engine along? Does the engine buck and surge as the gearbox takes the load of the pushing train?

Trains can be made to roll more freely with better trucks and wheelsets (I highly recommend metal wheels), and engine problems can often be solved with better motors or thrust washers in the gearbox. But be sure you can fix such problems or afford to pay someone who can before committing to a given gradient and minimum radius.

6 **The door between the author's basement and garage is temporarily blocked by two lift-out sections during operating sessions. The lower section just happened to be long enough to incorporate a deck-girder bridge in the proper geographic location. The top section will incorporate an under-cabinet fluorescent fixture that plugs into an adjacent fixture.**

Curvature

This applies to any track plan, but it bears repeating here: Choose your minimum radius with extreme care. It's far too easy to undershoot the mark.

When I designed the Allegheny Midland, I knew that 30" worked all right on Allen McClelland's original Virginian & Ohio, which I had visited on several occasions. He had some 36" curves, but the tighter ones looked okay to me. Of course, he was primarily using diesels, which tend to negotiate and look better on tighter curves than steam locomotives.

As I later backdated the railroad to the steam era, I found that Mikados (2-8-2s) and Berkshires (2-8-4s) handled the 30" curves okay as long as the drawbar pin was in the second hole, as did the 2-6-6-2s, but any thoughts of using 2-6-6-6 Alleghenies or even 2-8-8-0s were quickly forgotten when I saw how far the pilot and cab swung out on those curves. That wasn't really bad news, as a railroad the size of the AM probably wouldn't have had monster articulateds anyway. Then again, the Western Maryland and Clinchfield had 4-6-6-4s…

When it came time to design the NKP, I wanted the Berkshires to look good and operate well when the engine

The 42" minimum radius Bill Darnaby had adopted to accommodate big 4-8-2s on his Maumee Route without excessive overhang was also a good "visual" minimum for NKP 2-8-4s (top). The mountainous Allegheny Midland got by with 30" curves, fine for small articulateds (above right) but too sharp for bigger 4-6-6-4s (above left). **7**

8

A dozen waybill boxes for the 12-track east-end staging yard were moved from the open area on the fascia bordering this narrow stretch of aisle to a wider area to reduce congestion. The bill boxes are sold by Micro-Mark.

and tender were coupled a scale distance apart: 30" between the cab and tender. Bill Darnaby had the same objectives for his Illinois Central-based 4-8-2s, and he decided 42" was a good minimum. My own trials confirmed this choice, **7**.

Now consider this: Those 42" curves fit into the same basement that had barely accommodated the AM's 30" curves. The moral: Be careful what you aim at, as you're likely to hit it.

Design differences account for the larger minimum radius working out as well as it did, as the aisle width is similar. I think I now know how I could have bumped the AM minimum up to at least 36" – narrowing one yard would have helped – which goes to show that I probably gave up too easily back when I designed the AM. But it's also easier to justify tighter curves on a mountain railroad than on a prairie racetrack like the NKP.

Aisle width

Speaking of aisle width: If you study a number of John Armstrong's track plans, you'll find that he was in favor of more railroad and less aisle space. Some of his aisles are on the narrow side – downright skinny, in fact. The same was true on John's own O scale Canandaigua Southern. The railroad came first, and people could cope with what was left. That John was slim and trim may have accounted for his preferences.

You can see some of that philosophy reflected in Bill Darnaby's Maumee (see photo 1 in chapter 2). Most of the aisles are narrow but adequate, and the aisle near busy East Yard – the main classification yard – is sufficient. Where things get tight is at the east end of the railroad at Dacron, Ohio, which is on the top deck. Dacron is across the aisle from the turn-back curve at the end of the central peninsula, which marks the widest part of the railroad. When the Dacron yardmaster is standing on a stool to uncouple cars in the 66"-high yard, there isn't a lot of room to squeeze by. But crews do squeeze by, so operation wins another round over creature comforts.

In fact, operation on layouts with constricted areas becomes almost natural. Without even thinking about it, one waits at a wide area for an oncoming train crew to pass.

On the other hand, it's much nicer to operate on a layout with very wide aisles, just as it's nice to have a spacious crew lounge, carpeted floors, air-conditioning, nearby bathroom, refreshment counter with refrigerator, and so on. But the needs of the railroad should come first, assuming that basic creature comforts have been met.

Clear the aisles!

One of the major design goals for my NKP Third Sub layout was to eliminate control panels. If they project out into the aisle, they constrict people movements. If they're built into the fascia, crews tend to step back to see what control they're reaching for, which also constricts aisle movements. Even control panels built on sliding drawers take up aisle space when they're open.

I also wanted the jobs done by my crew members to emulate prototype jobs as closely as possible. This includes walking down to the point where a switch has to be thrown, as opposed to standing at a distant control panel and straining to see which way the switch points are lined.

(Let's again be sure we're on the same page regarding terminology. A turnout is the entire assembly of switch points, stock and closure rails, frog, and guardrails. A switch is the part of a turnout that can be moved. So you can't "throw a turnout"; it weighs far too much and is spiked down! You do "throw or "line" a switch, like professional railroaders do and say. Since the frog is not usually movable, it's part of the turnout, not the switch, and you therefore cannot have a "no. 8 switch" or a "high-speed switch.")

If you do not provide a place for yard and road crews to work, they will adorn your layout's surface with paperwork such as waybills and switch lists. It therefore pays to have boxes and small ledges where their paperwork can be stored when not being used. I used the three-slot boxes sold by Micro-Mark, which are designed to hold their (or Old Line Graphics) 2" x 4" four-cycle car cards and waybills.

The agent-operator's desk at Charleston (top) is located in a wide area of one main aisle. The Frankfort agent-operator's desk was relocated from a busy area across from Frankfort yard to the far end of that aisle (above).

9

I have moved several of these boxes to other locations when we discovered that they projected into busy aisles, **8**. A little more forethought would have seen these changes coming.

I also recently moved one of the two agent-operator desks that are located in the main aisles. Each agent-operator is responsible for the several towns and one classification yard in each major aisle, so his or her desk needs to be handy to those towns to enable train-order signals to be set, train orders to be hung from clips on the fascia, and empty-car needs for local industries tabulated and relayed to the yardmasters.

The operator's desk at Cayuga, Ind., which supports the west-end division point of Charleston, Ill., was located in a corner where the aisle is relatively wide, **9**, a good choice. But the desk at Linden, Ind., was clearly in harm's way, we discovered after several operating sessions. I therefore recently moved it to the far end of that aisle, which also serves busy Frankfort yard. It should be considerably quieter down there as

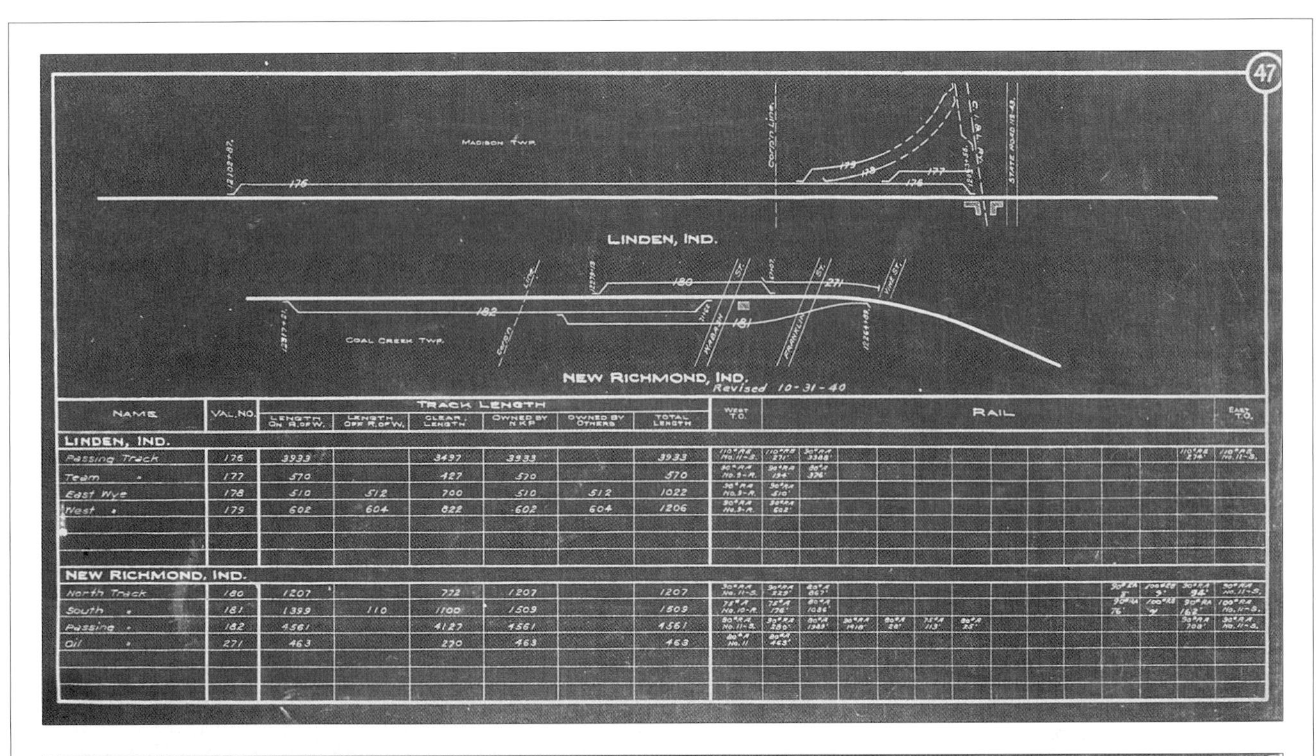

Name	Val. No.	Length on R. of W.	Length off R. of W.	Clear Length	Owned by N.K.P.	Owned by Others	Total Length
Linden, Ind.							
Passing Track	176	3933		3497	3933		3933
Team "	177	570		427	570		570
East Wye	178	510	512	700	510	512	1022
West "	179	602	604	822	602	604	1206
New Richmond, Ind.							
North Track	180	1207		772	1207		1207
South "	181	1399	110	1100	1509		1509
Passing "	182	4561		4127	4561		4561
Oil "	271	463		270	463		463

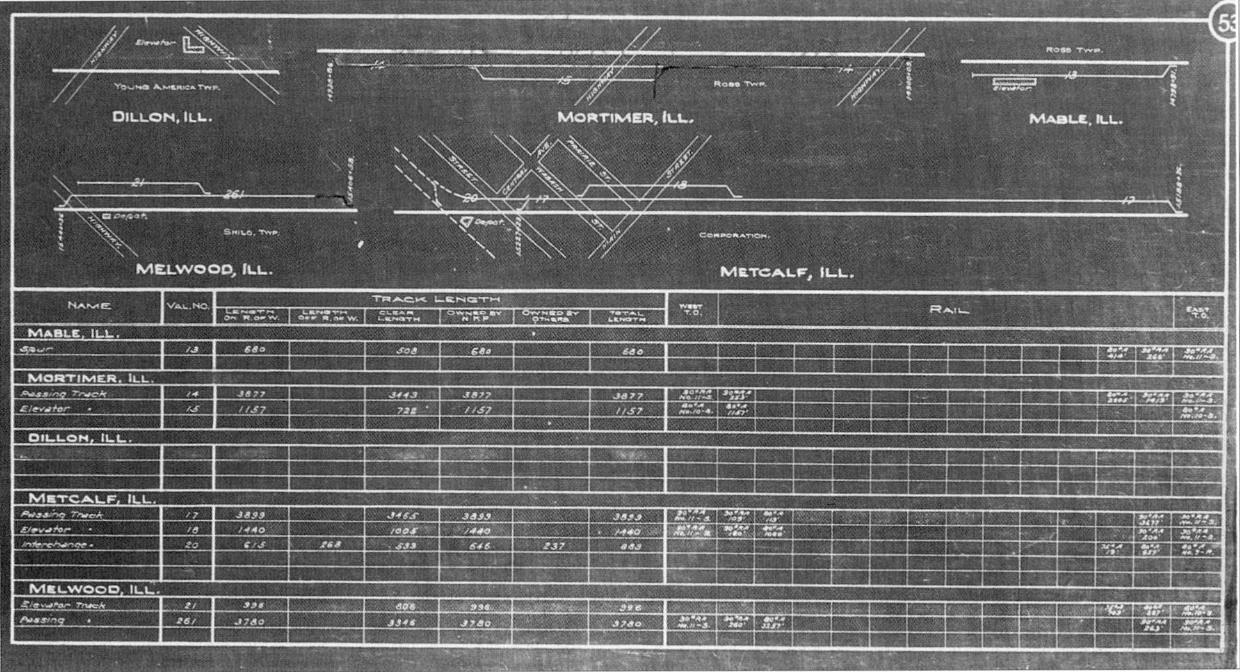

Name	Val. No.	Length on R. of W.	Length off R. of W.	Clear Length	Owned by N.K.P.	Owned by Others	Total Length
Mable, Ill.							
Spur	13	680		508	680		680
Mortimer, Ill.							
Passing Track	14	3877		3443	3877		3877
Elevator "	15	1157		722	1157		1157
Dillon, Ill.							
Metcalf, Ill.							
Passing Track	17	3899		3465	3899		3899
Elevator "	18	1440		1005	1440		1440
Interchange "	20	615	268	533	646	237	883
Melwood, Ill.							
Elevator Track	21	996		606	996		996
Passing "	261	3780		3346	3780		3780

These official Nickel Plate Road track diagrams for Linden, Ind. (top) and Metcalf, Ill. (above), illustrate the principle of "vertically paired towns." Note that the Monon interchange is to the east (right) at Linden, whereas the B&O interchange is at the west end of Metcalf. Even though Metcalf is located directly above Linden on the author's layout, crews working in both towns at the same time will be horizontally separated by ten feet or more.

10

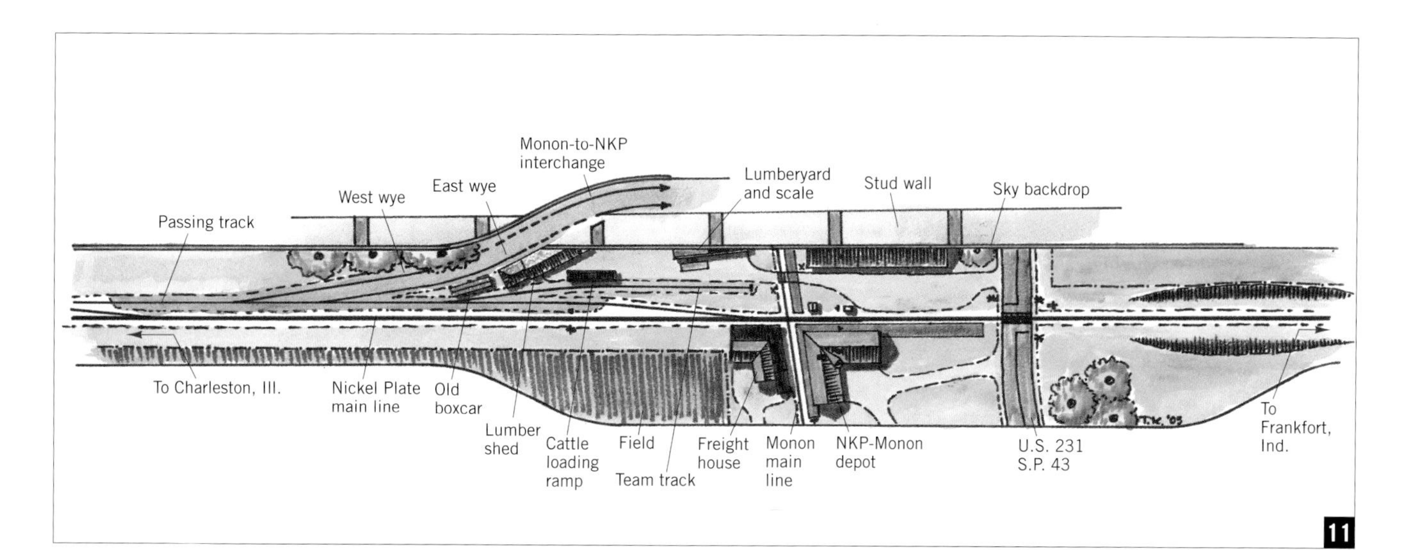

▲ The author's sketch of the track plan for Linden on his HO railroad shows how the NKP-Monon interchange resembles the wye used by the prototypes but then bends back to the right and parallels the stud wall that forms the backbone of the main peninsula. Monon deliveries to the NKP – some 30 cars per day – were automated using optical detectors from Circuitron.

well, making it easier for the operator to hear the dispatcher when he dictates train orders and messages.

Vertically paired towns

A major design consideration with a multi-deck layout is keeping track of where crew members will typically be standing. If you locate a town on the upper deck directly above a town on the lower deck, there is always the potential for crews working both towns at the same time to get in each other's way.

The easy way around this is to choose "vertically paired" towns. That is, the town on the lower deck might have its center of activity at the east end of the passing track, while the town on the upper deck has its focal point at the west end of its passing track.

Let me illustrate this principle by citing Linden, Ind., and Metcalf, Ill., on the NKP Third Sub. The NKP track diagrams, **10**, show the Monon crossing and interchange at Linden is at the right (east) end of the passing track, whereas the Baltimore & Ohio crossing and interchange is at the west end of the passing track in Metcalf.

Two photos: Judy Koester

During the construction of the upper deck on the author's layout, he realized that the top deck could not extend out to the aisle (as shown in the plan on p. 26) without blocking sightlines and even easy access to the rear tracks of Frankfort yard. The Humrick area was therefore cut back to the standard 16" width. It's quite a reach from the aisle at the east end (left), but there are no industries to work. The busier west end (right) is easier to reach, as Frankfort yard is much narrower there.

12

This means that crews working these two towns will tend to be 10 to 15 feet apart. Problem solved.

Extended interchange tracks

My sketch of Linden, **11**, shows how the interchange tracks between the Nickel Plate and Monon at Linden punch through a hole in the stud wall that serves as the spine for the central peninsula. Rather than being truncated at the sky backdrop, allowing perhaps three or four cars to be interchanged per day, the punch-through allowed me to run a pair of 30-car-long tracks on the far side of the wall.

Why make these tracks so long? The Monon and NKP interchanged 12,000 loads and presumably an equal number of empties during 1953, which averages out to a little more than 30 cars in and 30 cars out each day. Moreover, these freight cars were not delivered by either railroad all at once

13

Stephen Priest

By recessing the upper deck of his HO Santa Fe layout, Stephen Priest ensured that operators had sufficient access to a major yard on the lower deck. Lighting under the upper deck complements ceiling fixtures. Lenny Ohrnell has a good view of his train on the upper deck, but locating busy industrial trackage here would have been a mistake.

but rather throughout the course of the day as four or more freight trains stopped to work the interchange from either end.

I described how the Monon's delivery of cuts of cars can be automated in the September 2006 issue of *Model Railroader* using two off-the-shelf Circuitron modules. One is an optical detector that senses whether the NKP has picked up the last-delivered cut of cars and turns power back on to a hidden Monon engine if the interchange delivery track is clear. The other is a relay that actually switches the track power on and off.

The cross-section sketch on page 28 shows how I was able to tuck the Monon interchange tracks under the town of Veedersburg, Ind., on the opposite side of the peninsula. The gentle but constant westbound grade between towns (trackage in towns is level to make it easier to switch and spot cars without resorting to wheel chocks) provided just enough elevation gain for me to slip the interchange tracks under a thin layer of terrain.

Recessed upper decks

The original plans for Humrick, Ill., on my HO layout (page 28) called for the upper deck to extend out to the main aisle, directly above bustling Frankfort yard. But it quickly became apparent that this would block the Frankfort yardmaster's view of many westbound yard tracks.

I therefore narrowed the Humrick shelf to the standard 16". That makes it difficult to reach, especially on the right (east) end, where Frankfort yard is at its widest, **12**. Fortunately, the grain elevators once located here were gone by 1954. At Humrick's west end where cars are interchanged with the Milwaukee Road, Frankfort yard is much narrower, making it easy to uncouple cars by standing on a step stool.

A potential disadvantage of a recessed upper deck is that the lower-deck lighting cannot be mounted directly over the outer areas of the lower deck. Ceiling-mounted room lighting is normally sufficient to illuminate these areas, however.

Steven Priest recessed the upper deck above a yard on his HO Santa Fe layout, **13**. The photo, taken with available light, shows that the lower deck still receives adequate illumination from ceiling fixtures when complemented with lighting under the upper deck.

Brian Pate, whose HO and HOn3 layouts were covered in MRP 2007, reports that as upper-deck reach-in problems were encountered, he reduced the depth of the lower deck. He also installed access panels in the upper-deck fascia to ensure good access to switch-point and switch-stand mechanisms, and he has step stools located around the room, **14**.

Caveat: Adequate layout lighting remains the number one problem with many, perhaps most, multi-deck layout designs, so be sure you have this under control before construction starts.

Photoshop cures all

There's no question that having one or more upper decks causes some concerns when it comes to shooting model photos. Where a single-deck railroad could have an expansive sky or mountain-ride backdrop, a multi-deck layout has an upper deck.

Fortunately, in this day of digital photography and computer manipulation of images, such concerns are relatively easy to resolve through the use of Photoshop (page 90) or a similar software package. Is that cheating? I think once we decided to put electric motors in our steam engines, we had already conceded that point.

Crew size

Before leaving this discussion of up-front considerations for a multi-deck model railroad, let me again quote veteran modeler, operator, and professional railroader John Swanson. "One of the first considerations when planning a home layout is the size of the available operating crew. Something few of us consider is that, as we age, the construction and operating pool of local modelers may get smaller as formerly loyal crew members encounter physical ailments, move to warmer climes, or pass away. An overly ambitious plan may therefore prove more than we can handle.

"Conversely, you may have planned the railroad to accommodate a crew of eight to ten people. But as word gets around that there's a fun-to-operate railroad in the neighborhood, more and more modelers will want to participate. I built a new layout to accommodate 25 to 30 operators, but in the process of building a layout that could keep legions entertained, I may have lost sight of my original concept.

"I finally slowed the clock down to 2:1 and often annul switchers and branchline runs, and I can now get by with a crew of ten."

My own layout can accommodate up to 18 crew members, **15**, but it can also get crowded.

14

Dave Adams

▲ Several design features typical of multi-deck layouts are evident in this view of Brian Pate's HO and HOn3 layout: "Eye level" is relative to the viewer; ample shielded lighting is mandatory for both decks; and narrow rights-of-way increase the mainline run per square foot of floor space. Note the plastic guard in the aisle at lower right. Here Lori Neuman copes with her train on the upper deck while husband Seth works on the lower deck in the distance; that's owner Brian Pate at left.

15

Ted Pamperin

An operations-based model railroad – here the author's NKP Third Sub – tends to attract a crowd. That's doubly important on a multi-deck layout operated by timetable and train orders, as the number of available "jobs" roughly doubles as well.

"See" level vs. sea level

Here's a concluding thought that Doug Gurin brought to my attention: If you're modeling a railroad that ends at "tidewater" – an ocean port – or even at a major river port, should the harbor area be at the lowest point on the lower deck of the railroad? Will it look odd to climb up to water level, or to have a branch line that negotiates a helix and reappears below the harbor area?

Such "see level" transgressions, much like insisting on having east to your right (simulating looking north with the sun to your back), will bother some more than others.

What are our options?

Now that we've discussed some of the concerns we'll face if we embrace the multi-deck approach to layout design, in chapter 4 we'll take a closer look at our design options if we decide that more than one deck best needs our overall needs.

CHAPTER FOUR

Climbing between decks

The author's layout is a continuous spiral ascending to the west (left) except in towns, which were kept level to ease car-spotting chores while looking like their flatland prototypes. The grades between towns are generally less than one percent except here on notorious – on the model as on the prototype – Cayuga Hill, which still causes westbound High Speed Service freights to slow dramatically, and occasionally to "double the hill" across the Indiana-Illinois state line.

One size definitely does not fit all when it comes to adding one or more extra "stories" to a model railroad. My own railroad has two main decks, but there are three decks where the east- and west-end hidden staging yards reside below or above the two scenicked decks. The railroad is essentially a continuous corkscrew that slowly gains elevation as it progresses from east to west – except for yards and towns, where the railroad is dead level, **1**. But there are other ways to get between decks, and you can even build completely separate decks with or without a "virtual" connection if that best meets your needs.

Types of multi-deck layouts

There are two basic approaches to multi-deck layout design: the continuous spiral, which climbs steadily as it circles the room (this includes the so-called mushroom design), and the level-deck approach where the two decks are either connected by a helix or train elevator, or not connected at all. Let's examine each type.

The continuous climb

My layout (page 26) and Bill Darnaby's Maumee Route (photo 1 in chapter 1) are both examples of railroads that climb continuously in one direction. Mine climbs westbound, thus favoring superior eastbound trains; the Maumee climbs eastbound. With modest grades, neither climb-direction approach is problematic. Railroads with heavy tonnage predominantly in one direction, such as from coal mines to tidewater, should be designed to move tonnage downgrade.

My railroad is not built on a continuous grade; I kept the towns level to avoid having cars roll away when spotted at industries or on storage sidings, an important consideration with today's best free-rolling trucks. Bill's Maumee features some minor undulations to reflect the gently rolling terrain typical of northern Ohio.

My HO edition of the Nickel Plate's St. Louis line was designed to accommodate single steam locomotives on westbound (upgrade) trains. I hoped that by keeping most grades to 1 percent or less, a single Mikado or Berkshire would handle trains of 20 to 30 cars, a figure that turned out to be somewhat optimistic on two counts: achievable gradient and locomotive performance.

I'm now in the process of equipping the freight-car fleet with better trucks and wheelsets; Bill recommends InterMountain metal wheels in Accurail truck sideframes. Meanwhile, increased use of up to three first-generation diesels (Electro-Motive Division GP7s and Alco RS-3s) on freight trains or shorter trains behind steam has the situation under control on my layout.

▲ The summit of the westbound climb out of the Wabash Valley on the author's layout was originally planned for the corner in the left rear of this photo. He had to ease the grade by extending it another 30 feet so that steam engines could handle at least 20-car freights up Cayuga Hill.

Clearance between the lower and middle deck at Frankfort is a bare minimum – a 10" opening between the valance and fascia in what will be downtown Frankfort, Ind. It's tight but workable.

But let's consider the math: The grade between decks is directly related to the length of the mainline run per lap around the train room. I have roughly 250 feet of run per deck, and that means I could climb 1 percent times 250 feet (3,000 inches), or 30" by the time the main line reached the second deck.

But I can't employ all 250 feet of main line to climb between decks, as each town is level. Assuming a passing track of about 18 feet to accommodate 30-car trains plus caboose and locomotive(s) in three towns on the lower deck, that's 3 x 18 = 54 feet (648 inches) less mainline run than we assumed, so about 2,350 inches remain.

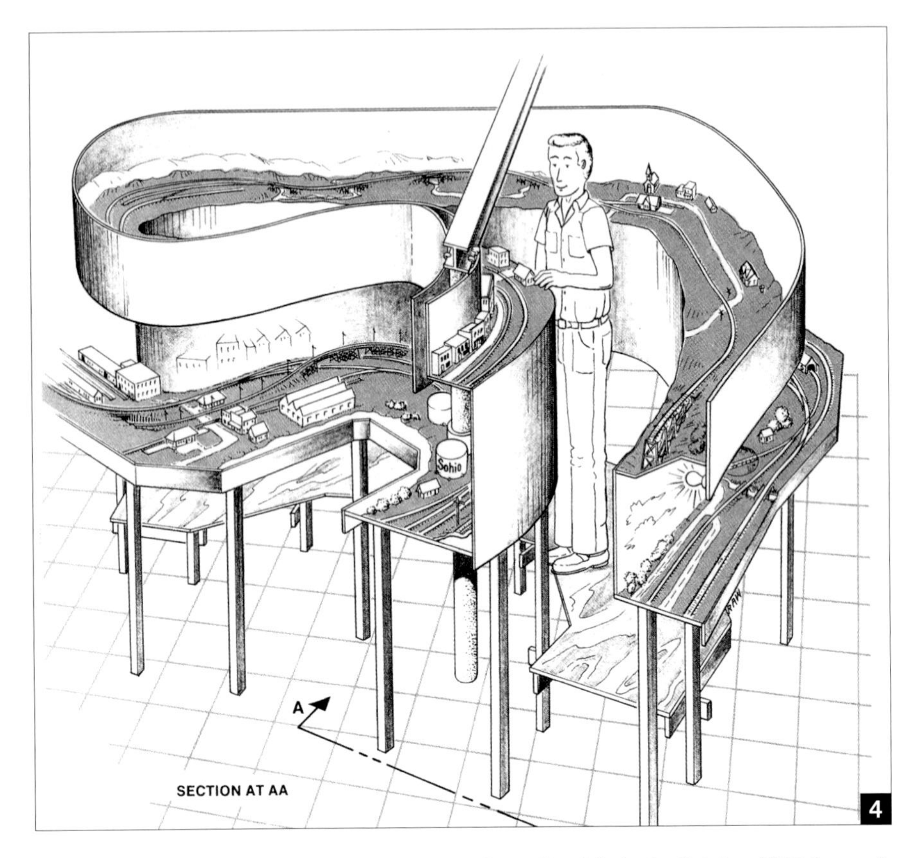

This sketch from John Armstrong's "Meet the mushroom" article in the October 1987 issue of *Model Railroader* shows how upper scenes overlap lower ones as the main line climbs. Note that the floors climb too.

And subtract the long classification yard at Frankfort, a loss of another 125 feet. So I actually had about 850 inches of mainline run in which to climb. At 1 percent, I could gain .01 x 850" = 8.5" between decks. Oops!

Fortunately, the NKP had a notorious grade west of Cayuga, Ind., so I could bump up the gradient to match the prototype's 1.29 percent there. The goal was to allow most 25-car trains to climb that hill without undue strain, but longer trains or poor train-handling skills would cause the train to stall, necessitating "doubling the hill" – cutting the train in half and taking it to the summit at Humrick, Ill., in two pieces.

I quickly discovered that neither I nor the NKP got to "vote" on the actual grade. It had to be determined by running test trains up the hill to see what was actually feasible. As a result, the summit was moved about 30 feet farther west to ease the grade, **2**. This actually aided clearance below the top-deck, west-end staging yard, but it hurt clearance above downtown Frankfort, **3**, as the middle-deck main didn't climb as rapidly.

Other mainline runs between towns had to have the grade steepened slightly to achieve a more desirable deck separation. My goal was about 15" to 16" from railhead to railhead, and I achieved that in most areas. But I continue to seek ways to improve the performance of the steam fleet with minimum car weights and better wheelsets.

Growing mushrooms

The so-called mushroom design is complex enough to warrant an entire book, let alone a chapter, but I'm going to skirt over it in one relatively short section. Instead of spending a great deal of time discussing what a mushroom is and how to build one here, I'll refer you to two of the leading experts on the subject, the late John Armstrong, **4**, and Joe Fugate. Joe sells a set of four DVDs (model-trains-video.com) that documents how he built his spectacular mushroom-style HO layout, which accurately depicts Southern Pacific's Siskiyou Lines set in the 1980s. His layout was featured in the January and February 1997 issues of *Model Railroader*. If you're even remotely considering the design and construction of a mushroom-style layout, reviewing Joe's helpful DVDs and Web site should be considered required homework.

The late Jerry Bellina was another practitioner of the mushroom method of layout design. The cross-section drawing of his layout in **5** and **6** should give you some sense of what his mushroom layout looked like. Henry Freeman described Jerry's innovative railroad in *Model Railroad Planning* 2003.

This approach holds great promise for the modeler who has high train-room ceilings. The need for extra ceiling height becomes apparent when you consider the primary objective, which is to retain the same floor-to-track elevation. As the railroad climbs enough for the second deck to fit above the lower deck, the floor also must rise about 12" to 18". Your train-room ceiling must therefore be high enough to accommodate that much floor rise.

Other prerequisites for those contemplating the construction of a mushroom layout are the ability to understand three-dimensional concepts presented in two-dimensional drawings and the knack of building a floor system that becomes higher in tune with the ever-climbing main line.

The basic idea is that one never sees the other deck along the spine of the railroad, as the railroad is viewed from only one side. This doesn't work for a perimeter wall, of course, so the goal is to build one or more peninsulas, a task made easier by the fact that each peninsula is only half as wide as a conventional peninsula owing to the overlapping upper- and lower-deck scenes.

Moreover, unlike with conventional multi-deck layouts, the modeler doesn't have to stoop to see lower deck scenes or stand on tip-toes to see upper ones, as the floor gets higher at the same rate that the railroad ascends.

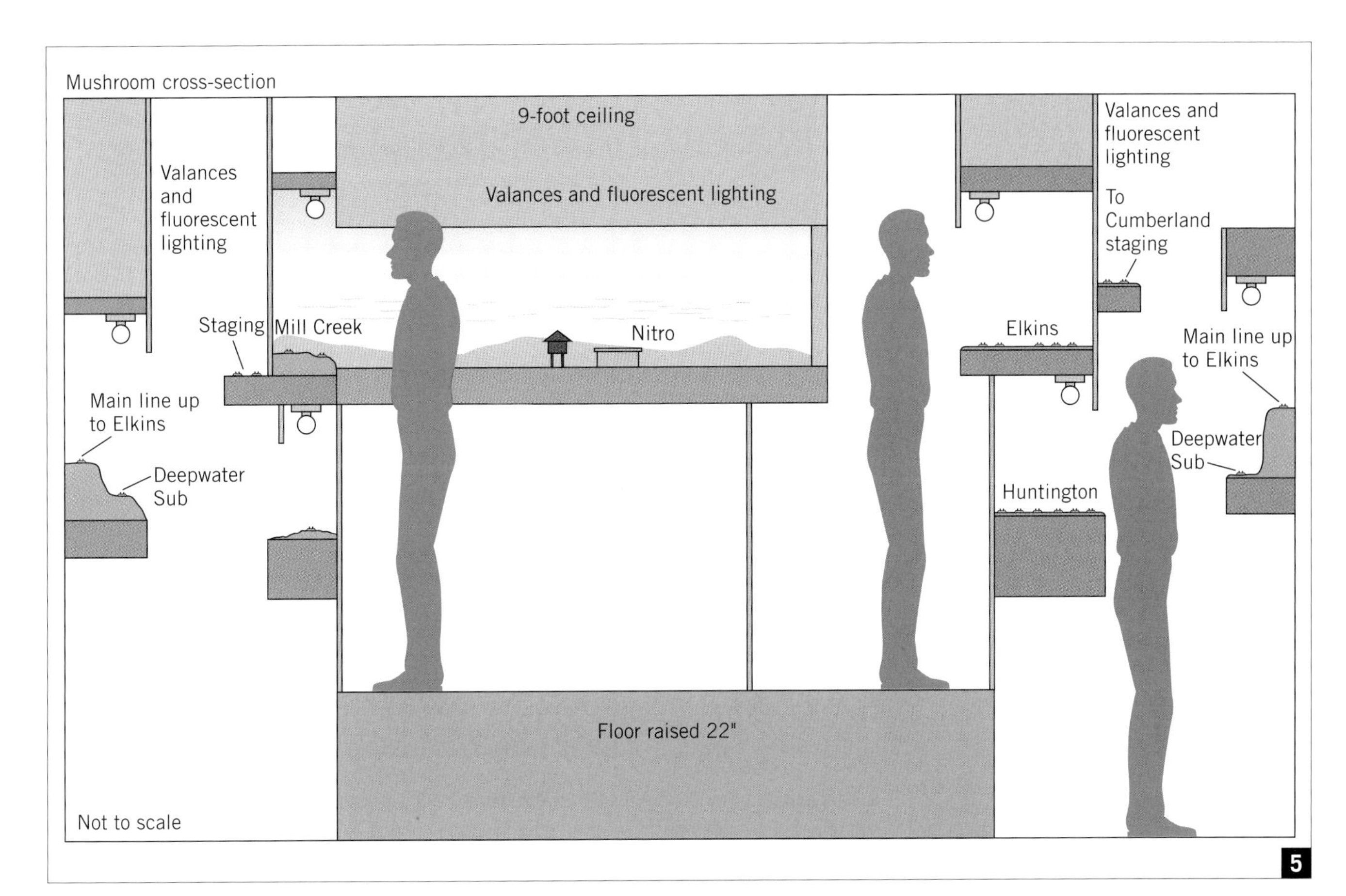

▲ **This cross-section drawing shows how the central area of Jerry Bellina's mushroom layout had a raised floor to accommodate upper-deck scenes. Lower-deck scenes were viewed from the main floor, and exterior-wall segments were of standard multi-deck design.**

► **Henry Freeman (foreground) runs a train through Huntington, W.Va. Elkins Yard is directly above this scene but accessible from a raised floor on the other side. Scott Dunlap's train is climbing to the upper deck.**

Paul Dolkos

Ceiling height

The ever-increasing floor height of a mushroom plan means that a high ceiling is required. Assuming you start out with a yard at, say, 40" above the floor and the second deck is 16" higher, that means the floor for the second deck is also about 16" above the base floor to retain the same 40" viewing height. Add around 7 feet (84") to that for adequate headroom and you have a floor to ceiling height of about 100", more than eight feet. And the railroad will probably still be climbing as it

Jim Vail

▲ **This helix on Guy Cantwell's HO layout has a lot of work to do: It moves trains between decks and also provides serial train staging (note second main) and reversing loops at various levels.**

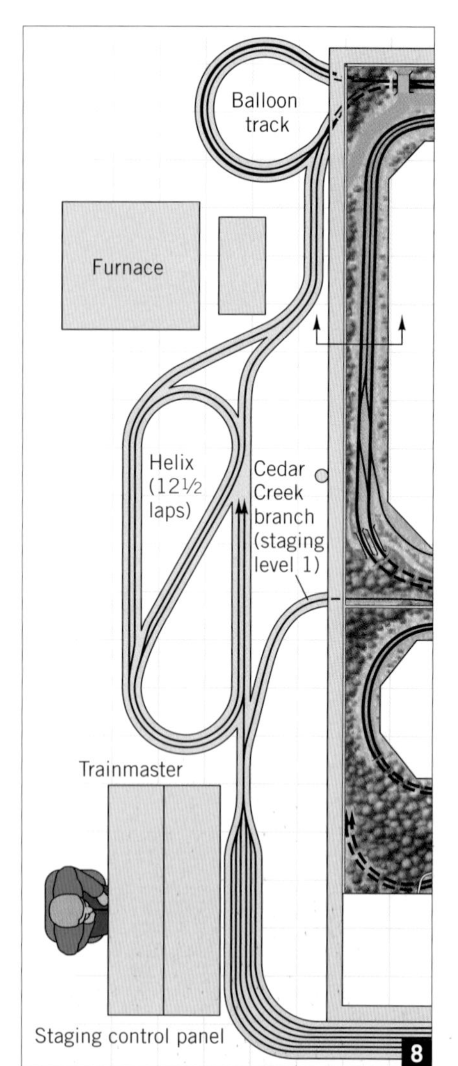

◄▼ **Multi-deck staging yards on Rich Weyand's heavily trafficked N scale Norfolk & Western layout (see *Model Railroad Planning* 2002) are connected with an oval-shaped helix, seen in the plan at left and in the photo below. This combination allows Rich to move loaded eastbound coal drags back to their points of origin.**

Rich Weyand

starts its second lap around the train room.

You can readily see why a typical eight-foot ceiling height is low for a mushroom design. Modelers who build new homes with their next layout in mind often specify an extra course or two of foundation blocks for just this reason.

René LaVoise's house, which dates to the 1920s, has only 82" of floor-to-ceiling clearance. He therefore employed a surface-mount ceiling system called Ceiling Max from Acoustic Ceiling Products (www.acpideas.com), available at Home Depot. Top rails are attached directly to the ceiling rafters, and standard 2 x 2-foot or 2 x 4-foot ceiling tiles are used. The bottom rail snaps into the top rail to retain the tiles. A similar product is made by CeilingLink (www.ceilinglink.com).

You can see photos of Thomas Cain's Ceiling Max installation at web.mac.com/tlc6383/Eastern_Illinois_Model_Railroad. Play the slide show that documents how he added the ceiling after adding furring strips to allow for pipes and ducts.

The helix

A spiral helix allows trains to corkscrew up or down between decks. A helix can be problematic, as trains tend to disappear into them for extended periods; they tend to have the railroad's ruling grade in a hard-to-see place; and they require a lot of floor space. (Doug Geiger sidestepped this problem by building his helix in a sealed box out in his garage; see MRP 1997, page 74.) A helix is also a bit tricky to build, but many techniques have been covered in the model railroad press and the model train article index at index.mrmag.com.

Jim Vail made a convincing argument on behalf of the helix in the 2008 edition of *Model Railroad Planning*, 7. As he pointed out, without a long linear run, climbing between decks can be almost impossible without using a helix. Among its other attributes is that it allows the modeler to build two absolutely level decks, as all the ups and downs take place within the helix.

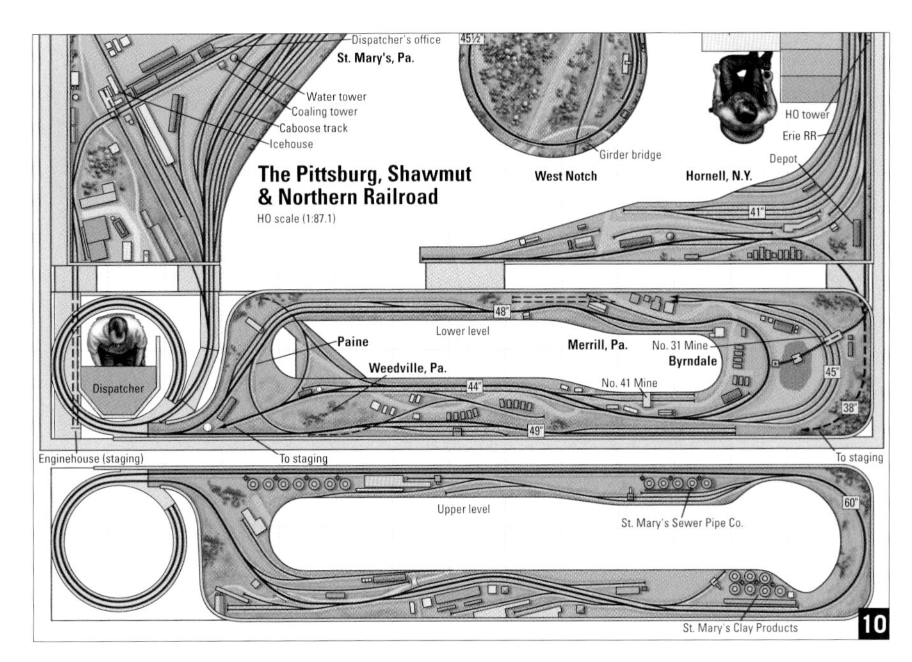

Perry Squier utilizes a double-track helix to allow trains to climb on the main line from a valley to an upper-deck summit, where the coal branch descends once again to another valley on the lower deck.

▲ When a train traverses a helix, it may change apparent geographical direction – east is now to the left. This may be an advantage if the railroad changes sides of a river (here on the Arkansas & Missouri in April 1988) and causes the viewer – who is usually "standing in the river" – to be looking south instead of north.

► In his primer on helix design and usage in *Model Railroad Planning* 1997, Doug Gurin showed how a helix can be manipulated to maintain a train's apparent geographical direction – usually east to the right.

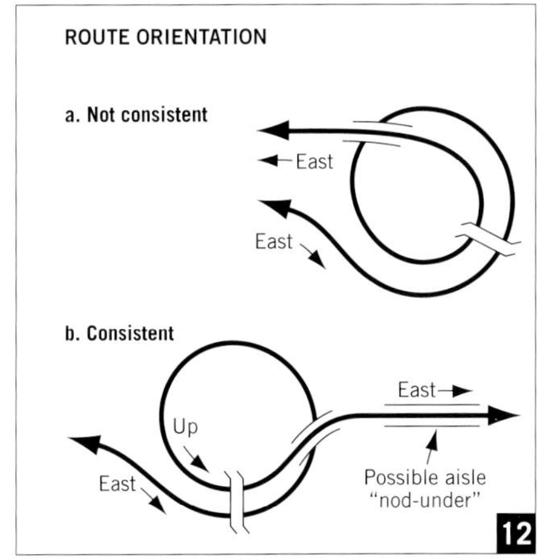

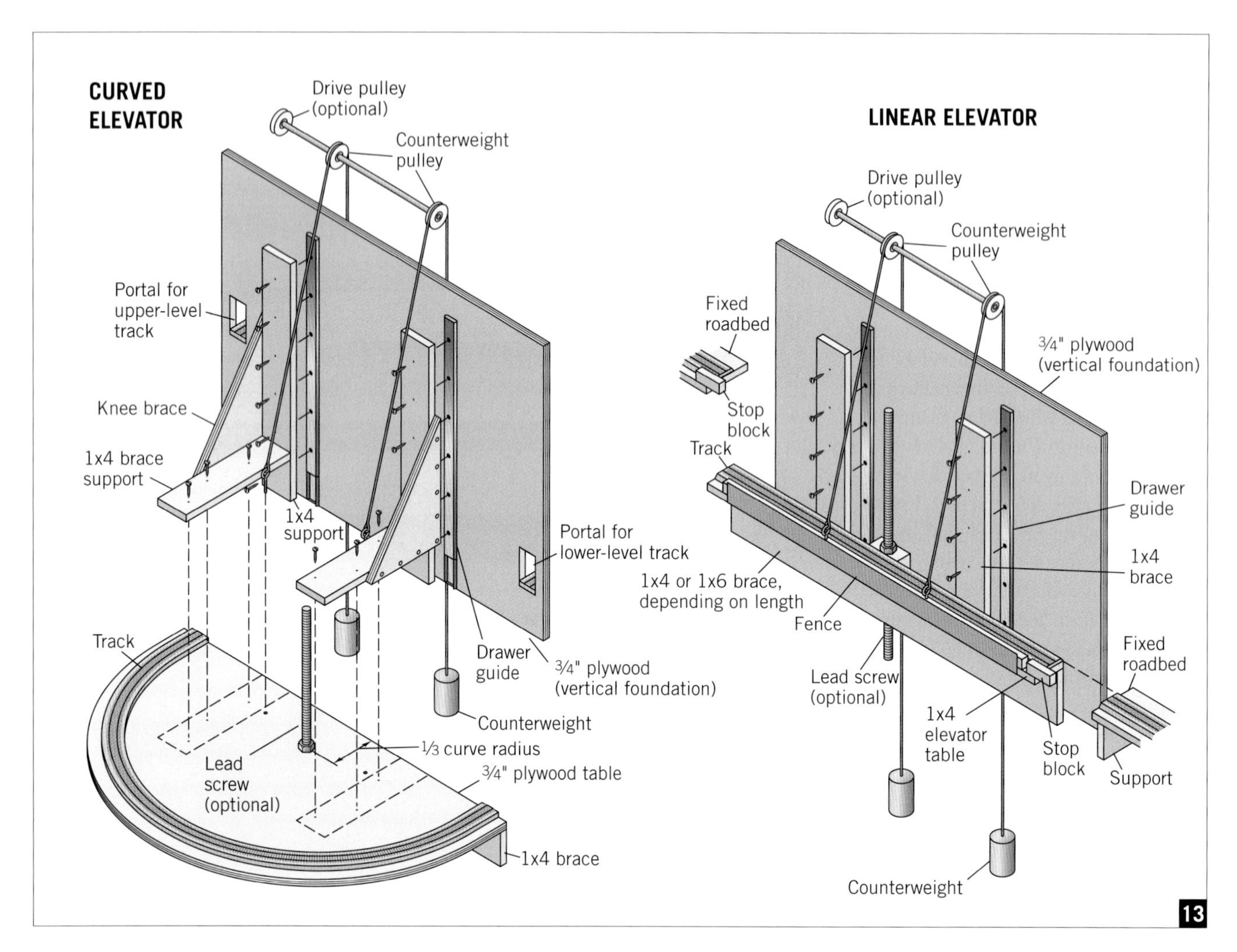

The train elevator, which moves entire (but short) trains between decks, need not be located on tangent track, as Dick Roberts explained in *Model Railroad Planning* 2001. Commercial train storage elevators can be adapted to this use.

Doug Gurin's thorough overview, "A primer on helix design," appeared in the 1997 issue of *Model Railroad Planning*. Among his many ideas was a suggestion to use the hidden area inside a helix for an isolated branch line, a technique that can also be applied to the inside of a peninsula's turn-back curve.

Rich Weyand built one of the more complex helix installations that I can recall, **8** and **9**, but it did the job for the heavy-duty mainline traffic on his N scale Norfolk & Western layout, as he described in MRP 2002.

Double-tracking a helix is often a good idea, perhaps by simply extending a passing track that begins and ends outside the portals to the helix. This allows more fluid train movements over what is almost certain to be the slowest-moving section of the railroad. At the very least, as with any major grade, it's a good idea to have several passing tracks at both the top and bottom of the climb, as this is where back-ups are most likely to occur.

Perry Squier designed a clever double-track helix at the beginning of an extension to his Pittsburg, Shawmut & Northern HO layout, but the two tracks are not connected, **10**. One track of the helix is used to climb from the main yard at St. Marys, Pa., to the summit at Paine Junction. The railroad then loops around a separate room on the upper deck before descending on the other helix track to reach coal tipples down in the valley on the lower deck.

One apparent disadvantage – the east-west direction of a train often reverses as it leaves the helix – can sometimes be an advantage: The viewing direction flops from north to south, simulating what happens when a railroad crosses a river and runs on the opposite bank, **11**. Often, paralleling highways do just the opposite, which reverses the viewing direction from an automobile following a train. But, as Doug Gurin pointed out in MRP 1997, this direction reversal can be avoided with careful design, **12**.

Train elevators

Dick Roberts described an innovative train elevator, **13**, in MRP 2001 ("Going up!"). He uses the elevator to move entire trains between decks on his two-deck HO scale Nevada County RR. If you're thinking that the moving track has to be tangent (straight), think again!

To avoid building such a complex system, you might investigate the roll-on, roll-off Ro-Ro Elevator Display (www.ro-ro.net). This product may also offer a way to locate staging tracks vertically against a wall instead of in a horizontal yard. Ro-Ro's Walt Kiefer reports that several customers have used the elevator to move trains between decks. The maximum height gain or loss is 25⅛" at 3⅛" intervals with consistent alignment after the initial setup.

Separate, unconnected decks

In the 2008 issue of MRP, Byron Henderson described how he would model Oahu Ry. & Land Co.'s sprawling three-foot-gauge system in Hawaii. He used an 11'-10" by 11'-11" spare bedroom as a home for his innovative plan, which comprised three decks.

The novel aspect of his plan is that none of the decks are physically connected. As shown in **14**, the decks are connected virtually, although any one of them could be built as a stand-alone layout, as could any two of them. If the decks are operated as a contiguous railroad, then a train that leaves one deck by entering a staging yard could "reappear" as a similar train emerging from a staging yard on the deck above or below it.

Jim Diaz models the Western Pacific around the famous Keddie Wye area in California, but he built a logging line on a separate deck that is not connected to the main railroad. This gave him the opportunity to do some modeling of a very different nature to the WP's busy main line.

I recently heard of a modeler that couldn't choose between two modeling scales and therefore built a railroad in HO on one deck and N on the other.

Choices, always choices

So there you have it: Connect decks by a continuous corkscrew climb, connect them with a helix or train elevator, connect them with movable interchange tracks, connect them virtually through staging, or don't connect them at all.

If you're still not sure you want or can manage the construction of a multi-deck layout, I recommend that you simply build one deck and connect both ends to staging or a fiddle yard. Once that is under control, then consider adding another deck. Just be sure you provide for the easy addition of another deck as you build the first one. In fact, this approach is far more common than you might assume.

Even though we now have a handle on how we might move trains physically or virtually between decks, we need to consider how high the decks should be. We'll do just that in chapter 5.

Kent W. Cochrane

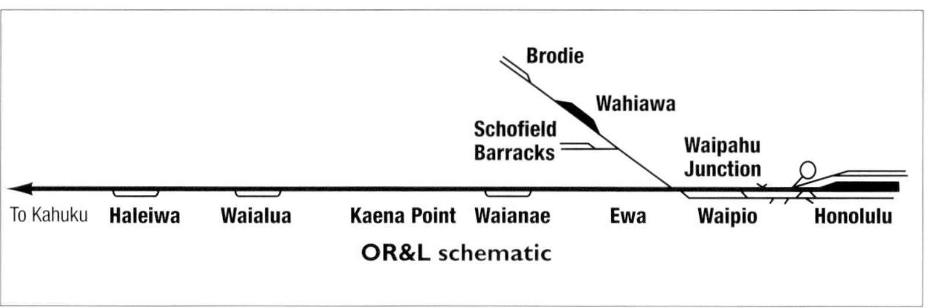

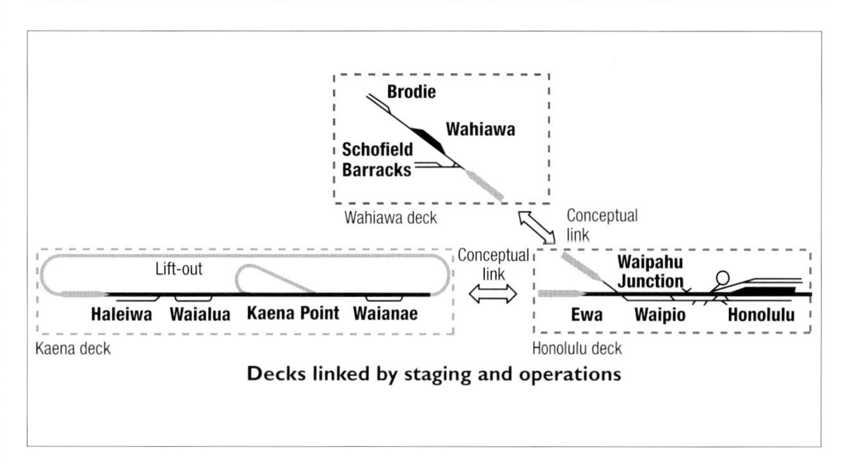

Decks that represent different parts of the same railroad can be connected "virtually" rather than physically, as Byron Henderson explained in MRP 2008 using Hawaii's spectacular Oahu Ry. as an example. Each deck could also represent a completely different railroad – perhaps the two railroads that cross at grade in a town modeled on both decks.

14

CHAPTER FIVE

Height compromises

The Nickel Plate located its east- and westbound yards on either side of the main line at Frankfort, which translates to a long reach-in for the westbound yardmaster on the author's HO layout. A step along the skirting makes the reach easier to handle.

There's no way that a roomful of modelers of different stature are ever going to agree on the "best" elevation for a model railroad. What is eye level for a six-footer is going to be rather high for a five-footer. Even if we were all the same height, there still wouldn't be an optimum height, as what works well for realistic viewing is probably too high for easy construction, maintenance, and operation. Try reaching across a dozen yard tracks to uncouple a car when the yard is six, five, or even less than four feet above the floor, **1**. And, even if we find some way around that impasse, the instant we decide to add a second deck means that one of the two decks, and probably both of them, are not going to be at what we'd consider an ideal height.

Testing, testing ...

The only way to become comfortable with layout height decisions in advance of layout construction is to build some full-size mock-ups of various benchwork height and depth combinations. Just as it's critical to test-run locomotives with typical trains both up and down the grades you plan to use (will the engines you want to use pull a desired train up the desired grade, or come back down without surging or sliding?), it's equally important to see for yourself what a given layout height looks like, **2**.

In *Model Railroad Planning* 2006, Steve King recommended using a typical bookcase that employs removable pegs to change shelf height to test various benchwork elevations, **3**. The string shows how the track elevation increases and then decreases before climbing again at various stations along the main line as well as on branch lines.

Steve placed rolling stock and structure models on each shelf so he could better visualize how the railroad would look at each major elevation. He could also try reaching and seeing over foreground models to uncouple cars on a track toward the rear of each scene.

Most bookcase shelves are around a foot deep, which is acceptable for many N scale scenes and some larger-scale scenes. My NKP layout uses 16"-wide benchwork and subroadbed, **4**, except in classification yards and in towns where an aisle-side, L-shaped depot or industry requires a wider footprint. In the latter towns, I typically added an 8"-wide (24" total width) "bump out," **5**, to accommodate the structure.

Busy Frankfort yard, with its separate, side-by-side east- and westbound yards, ranges from 38" to 47" wide, **6**, the latter requiring a raised step along the skirting, **1**, so the westbound yardmaster can more easily reach in to uncouple cars. Charleston yard is only 24" wide, but its 68½" height required a raised floor and step, **7**.

What others have done

Ken McCorry has one of the largest home layouts ever built, **8**. His lowest main deck is 36" above the floor, and the upper main deck is at 54", a spacing of 18". However, he has staging yards and secondary lines that are only 26", and Keating Summit towers above the floor at 75".

When Ken recently reflected on his layout's design, he commented that he "would probably increase the spread between the main decks by 4" to make access easier. As the depth of a scene increases, the height between them should also increase to allow room for your body to fit between levels."

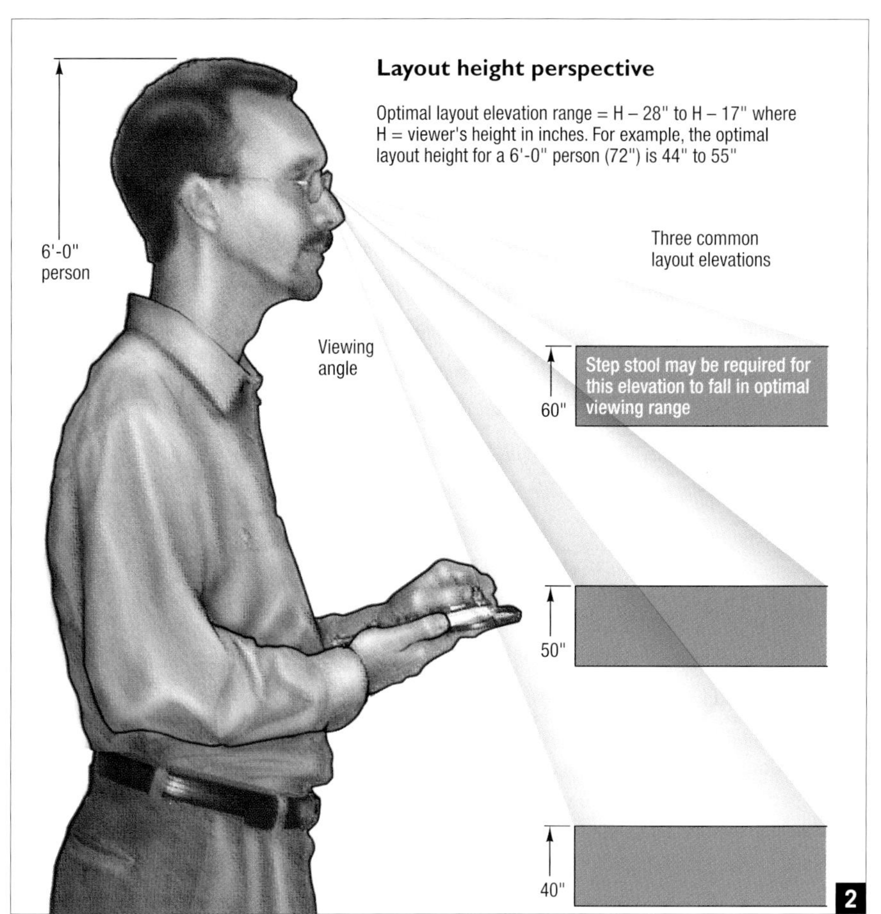

Rick Johnson

▲ Three common layout elevations from 40 to 60 inches show that a step stool may be required even for a six-footer to view and operate a layout within an optimal height range. The drawing also hints at lower-deck viewing limitations caused by upper decks.

Steve King

► Steve King used a bookcase, tape, and string to visualize the rise and fall of his N scale Virginia Midland layout. By moving shelves to proposed deck heights, he could determine in advance whether this would create viewing or operating restrictions.

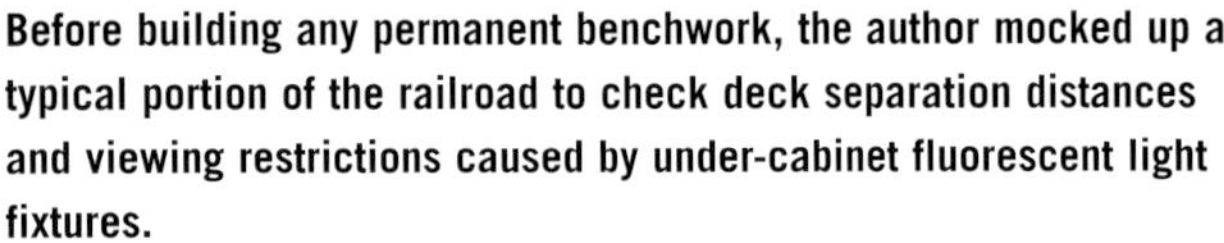

Before building any permanent benchwork, the author mocked up a typical portion of the railroad to check deck separation distances and viewing restrictions caused by under-cabinet fluorescent light fixtures.

To accommodate L-shaped depots on the aisle side of the NKP main in two towns, the author broadened the typical 16"-wide benchwork to 24". A similar bump-out on the upper deck was cause for concern, but so far no one has bumped into or raised up under it.

John Swanson is another builder of a large multi-deck home layout. At 6'-2", "high benchwork" for John starts around 65". But shorter or younger operators may find benchwork above 52" to be a challenge, he observes. John believes that "a sit-down lower yard at about 28" is a good place to start," **9**. Towns that can be worked sitting down can be in the 28" to 36" height range, he suggests, noting that he has several operators who have various infirmities and hence prefer to work sitting down.

"I also have an upper-deck staging yard at 76" that you can walk under. The higher end of the railroad requires step stools for access and a platform for the yardmaster."

Yard design

Don't make the mistake of assuming that the yardmaster has to be able to see reporting marks and car numbers of cars on each track. He or she should have sorted the cars' waybills (or updated a switch list) as the cars were classified into each yard track, then stored those waybills, in order, in a separate bill box for each track. If a car goes missing, rest assured it will eventually turn up. If for any reason the order of cars on a particular track becomes unclear, simply pull the entire track out onto the ladder so the cars' reporting marks and numbers will be visible and rearrange the waybills to match. Going on a treasure hunt for a particular car is not efficient.

Aisle-side industries

Let's again briefly digress from our multi-deck discussion to ensure that we agree on terminology. An "aisle-side" industry is one located between the main line and the fascia (aisle). This is a bit risky in that it puts the structures and cars serving it close to the aisle, where elbows and derailments can cause undesirable outcomes. The taller the railroad is at that location, the more protected the models will be.

I strive to keep anything frangible at least 6" back from the fascia. That includes not only the main line but also aisle-side sidings and spurs. This is obviously not possible where a wye track or industrial spur curves off the edge of the benchwork.

On the plus side, anything that momentarily blocks one's view of a passing train divides a scene into two mini-scenes, which is good in that one "event" is converted into two separate events – the train approaching the view block, and the train reappearing from behind the view block. Such view blocks can be short tunnels, **10**, high hills, groves of trees, tall structures, **11**, or a railroad or highway overpass.

A "Bellina-drop," named for pioneering mushroom layout builder Jerry Bellina (see photo 6 in chapter 9), accomplishes the same thing while passively suggesting to crew members that they not stand in an undesirable place, such as a constricted aisle at the end of a peninsula.

Rounded-corner panels of Plexiglas or other clear acrylic are a common way to protect structures or signals that are located close to the aisle, **12**.

Viewing angle with height

As shown in **2** and **4**, the height of a scene compared to the viewer's eye level has a huge effect on what he or she can see. On a single-deck railroad,

Frankfort's westbound yard is on the north side of the main line, which requires the westbound yardmaster to reach over the engine ready tracks and eastbound yard lead when working the east-end ladder.

only the valance, if any, and tall foreground structures can restrict one's view of even the deepest scene. But the combination fascia/valance of a higher deck on a multi-deck railroad may severely restrict one's view of the lower deck, especially if the benchwork and scenery extend well back from the aisle.

The higher the decks, the closer adjacent decks can be to one another. Imagine a scene with the deck being viewed right at eye level, **13**. Clearly, you could look into, but not down on, a scene of almost any depth, even if the decks were only a foot or less apart. But if that scene were located well below eye level, as on my layout in downtown Frankfort, Ind., **3**, one's view is considerably more restricted.

At the suggestion of Karen Parker, I bowed the middle-deck fascia/valance well in from the lower-deck fascia at the east end of downtown Frankfort, thus affording a much better view of a spur back into the local power and light plant, **14**.

Construction concerns

It's not difficult to work between decks when they're around 16" apart, but anything much tighter that than offers head-knocking challenges. I thought I could get away with some tight separations on my NKP, as I wasn't planning to handlay all of the track,

When completed, Charleston yard will have a main line, seven classification tracks, and an engine lead in its modest 24" width. However, its 68½" elevation required a raised floor and step to ensure easy uncoupling access.

Two photos: Brad Bower

Ken McCorry managed to keep the interior area at the end of a main peninsula open by using cantilevered benchwork that extends out four feet (above left). The dark bottom deck is staging. Ken used black cloth for the valance and skirting. At right, Ken stands alongside his layout at a point where there are four decks with staging on decks 1 and 3. Note the Wilkes Barre staging yard's train paperwork storage – one slot per track.

but I soon discovered that (1) I did have to handlay a wye turnout in one tight location, and (2) even commercial track requires a lot of fussing to get it aligned properly and to adjust it from time to time.

Fortunately, I can see without my glasses extremely well at very close distances (Mother Nature's sorry-about-that to those who have astigmatism), so shoving my head into such tight confines is uncomfortable but not impossible to deal with.

So what should I have done instead? Probably nothing could have been done. The lowermost part of the railroad, Frankfort yard, was set at 43" so that the west end of the railroad would be high enough to clear a knee wall I installed to steal the rear nine feet of our garage for the railroad; the knee, **15** allows the nose of our car to fit into the garage, if just barely. (It was built to meet the fire code, I should add.)

That plus the average grade as the railroad gradually climbed westbound from Frankfort to Charleston, Ill., determined the amount of clearance between decks in downtown Frankfort – a measly foot – as the railroad began its second swing around the basement perimeter and central peninsula.

As we previously discussed, I couldn't increase the grade, since I wanted to run up to 30-car trains behind 2-8-4s. And the Nickel Plate didn't double-head steam engines except to balance crews and power between division points. Had I modeled the diesel era, life would have been much simpler: I could have added another unit if need be.

(Keep such considerations in mind as you choose an era to model. If you don't already have a good-running steam fleet, I highly recommend setting your railroad in the diesel era and then gradually working back to the steam era one engine at a time. You'll have challenges aplenty trying to debug the railroad and its control system without having simultaneously to wrestle a cranky steam fleet into submission.)

Another problem with increasing a grade to achieve greater separation is that this also increases the height of the other end of the railroad. My calculations showed that the railroad would climb from 43" at Frankfort to about 68½" at Charleston, so I certainly wasn't looking for ways to make the west end even higher. Efforts to make the high end somewhat lower proved unsuccessful.

A helix would have allowed me to keep the decks at constant elevations, but I choose to emulate the design of Bill Darnaby's Maumee Route with its continuous climb. As discussed in chapter 4, a helix can be an effective design tool, but it also has its liabilities, including large footprint, and it often hosts the railroad's ruling grade. Having long trains get out of the yard just fine but then stall in a hidden coil is not something I wanted to risk.

Had my layout room been too small to accommodate a run that was long enough to achieve the desired deck separation, however, I would have installed a helix.

Elevating the floor

After Frank Hodina drew a much better track plan than the one I concocted, I studied it for potential problems or opportunities to enhance it. One of Frank's improvements was to move the Charleston, Ill., yardmaster's working area into an alcove between the yard and the basement's perimeter wall (pages 26 and 28), which allowed me to raise the floor by screwing ¾" plywood atop 2 x 10 floor joists spaced the standard 16" apart, **7**. Raising the floor even higher would have caused headroom problems for anyone taller than my 6'-3" height.

Once this elevated floor was built, I found that I could reach all seven tracks in the 24"-wide yard, but just barely. The floor reduced the actual 68½" yard height to a more manageable 60", and I found I could reduce the height even more by building an 8"-high step along the length of the yard, which shows clearly in **7**. I kept the front edge of this step open so the yardmaster's feet could fit under it during normal yard operations. He or she can then gain a little extra reach by mounting this step with one hand holding onto the handrail and the other uncoupling cars.

Reach-in concerns

The step at Charleston proved so successful that I copied it along the edge of Frankfort yard to give the westbound yardmaster better access to the busy east ladder area, **1**.

Why is a step needed at Frankfort, which is only 43" high? No step is needed for the eastbound yard, which is adjacent to the aisle. But the westbound yard tracks are on the north (most distant) side of the main line through Frankfort (see the track plan on page 26), and their center lines range from 24" to 38" back from the aisle. That's quite a reach – well beyond the recommended maximum of 30" – and I assumed a footstool would be adequate to increase the yard crew's reach as they uncoupled cars. The need for the westbound yardmaster to lean forward puts him or her in a precarious position at best; the stool could easily flip out from under his or her feet. The carpeted elevated step solved that problem.

Why not use permanent magnets or electromagnets to uncouple cars on the more distant tracks? I'm not a fan of permanent magnets between the rails, especially on busy yard tracks or main lines, as they cause unwanted couplings almost as often as they allow cars to be uncoupled as desired.

Two photos: John Swanson

Many modelers are following John Swanson's example of "sit-down yards" at 28" to 30" for the lower deck (above left and right), which lowers the elevation of the upper deck(s). He has also recessed the upper deck in some areas (above left) to create a more-open urban environment.

9

10

Charlie Comstock

▲ The two narrow ridges penetrated by the railroad separate one long scene into three different "events" on Charlie Comstock's HO railroad set in the Pacific Northwest. Tall structures, hills, or even groves of trees between the main line and aisle accomplish the same thing. Note the tunnel access port in the fascia. That's Paul Mack running the freight.

▼ The large passenger station and street spanning the passenger terminal on Bruce Chubb's Sunset Valley Oregon System divides one long scene into two more-intensive ones. Modeling this scene in a single-deck area also accents an important feature of the railroad while avoiding upper-deck accessibility problems.

11

Bruce Chubb

Electromagnets or movable (up/down or sideways) permanent magnets work much better, but they require accurate spotting of cars and marking the magnet's position. I therefore prefer that crews work with some sort of skewer or pick. Raising one end of a car to effect uncoupling is not acceptable for a whole host of reasons ranging from greasy fingerprints on cars to detail damage.

Ceiling considerations

A finished ceiling is beneficial in that it enhances the look of the railroad room while preventing dust and debris sifting down onto the railroad. But it may be more difficult to fish wire between recessed lighting fixtures or reach the plumbing above a basement layout room, and the added thickness of the ceiling system reduces headroom.

A drop ceiling makes it easier to gain access to wiring and plumbing

Tommy Holt

▲ **Tommy Holt uses clear acrylic shields fastened to the fascia at locations where elbows or track cleaning efforts are likely to damage delicate lineside structures such as signals.**

▼ **Seth Neumann has an excellent view of, and access to, a yard on the lower deck of Brian Pate's HO and HOn3 layout. He has a more-realistic, eye-level view of the narrower top deck, but working a busy yard at that height would require the use of a step stool. Recessing the upper deck enhanced the view of the lower deck and accommodated the tall tipple.**

Dave Adams

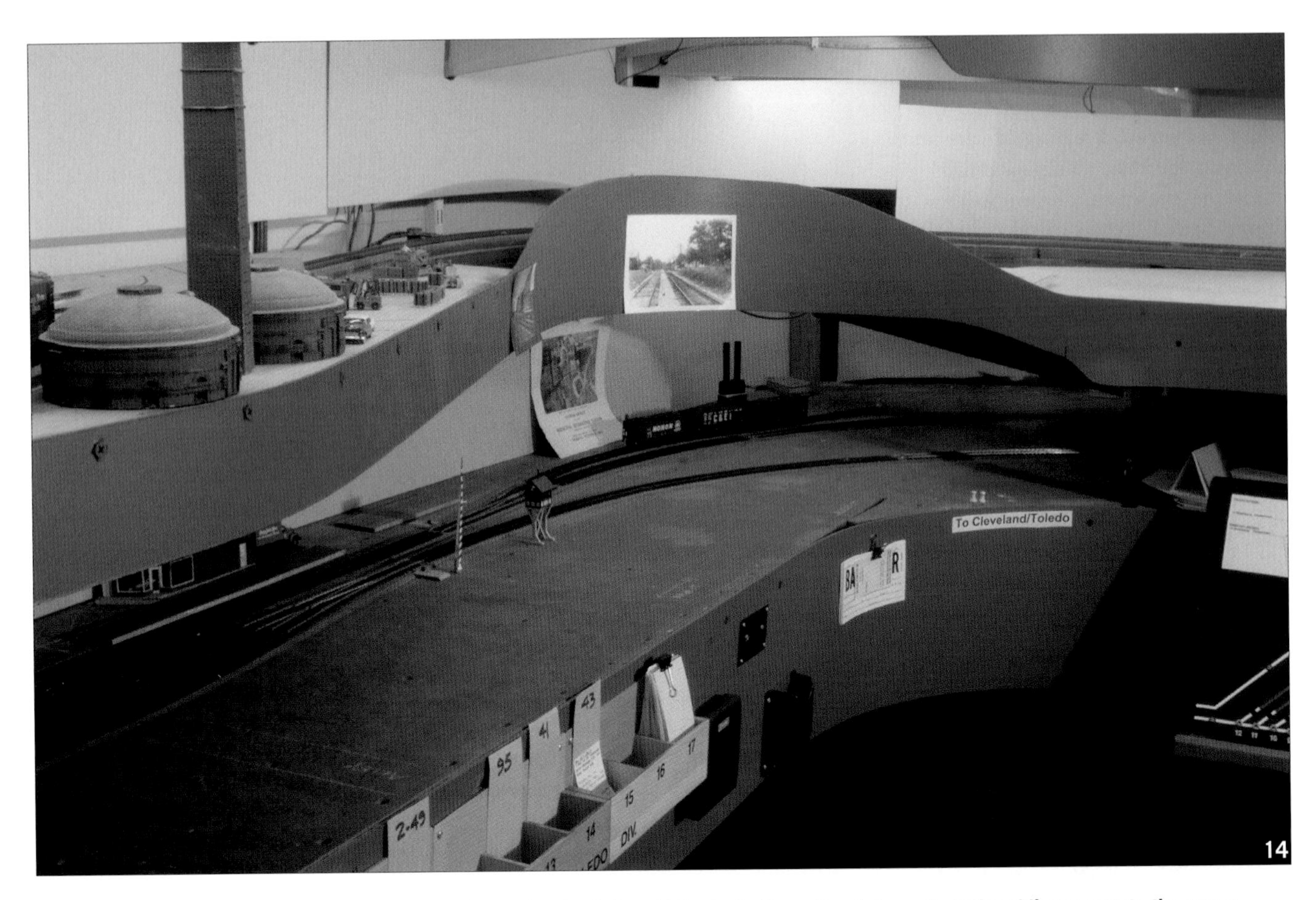

After following Karen Parker's suggestion to bow in the fascia of the middle deck, the author felt comfortable adding a spur to the power company in a previously hard-to-reach corner of the lower deck at the east end of Frankfort, Ind.

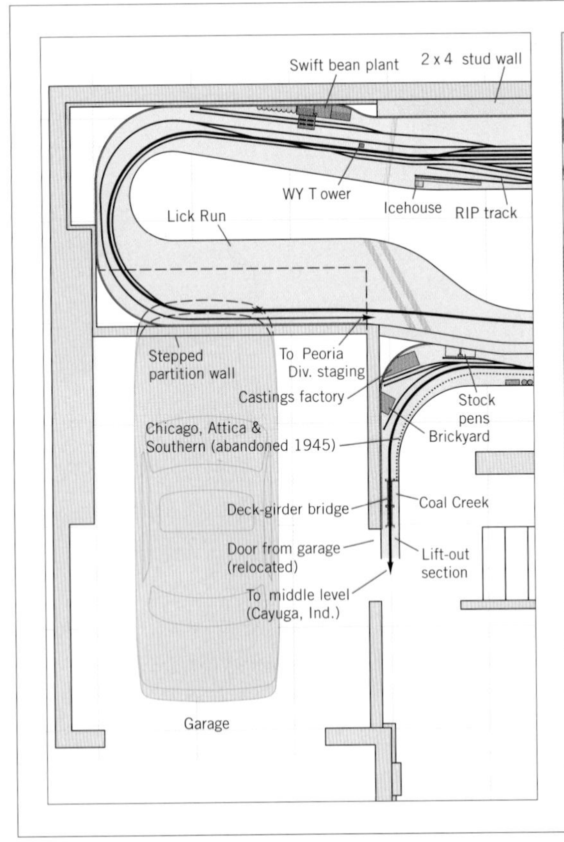

The rear part of the author's attached garage was "re-purposed" by building a knee wall to separate the garage from the basement, allowing the main line to swing out over the nose of the car. The extra mainline run was about 30 feet per deck, or a scale mile in HO!

15

above the tiles, but ceilings of this type may reduce the headroom by several inches.

I started with a partially finished basement and screwed 4 x 8-foot sheets of wallboard to the ceiling, then taped and spackled the joints. This minimized height loss compared to a drop ceiling, which proved very important when I later replaced the original single-deck Allegheny Midland layout with the multi-deck Nickel Plate Road. The continuous-spiral nature of the NKP's design means that one end (the west end) of the railroad is considerably higher than the lower end – it climbs from 43" at Frankfort, Ind., to about 69" at Charleston, Ill., a gain of more than two feet.

As we have already discussed, the extreme height of the west end led me to build an elevated floor for the yardmaster's and engine hostler's alcove at Charleston, and the amount I could raise the floor was dictated by the existing ceiling height. Had I employed a drop ceiling, headroom would have been insufficient.

Check the Web or building supply centers for new types of ceilings that are considerably thinner yet still removable.

16 The author covered the basement's concrete floor with carpet tiles to provide a comfortable operating environment. Each 22" x 23" tile has a non-skid foam backing and is not glued down, so it can be picked up if it gets wet or is damaged.

Carpet tile

Let me digress once again from issues related only to multi-deck layouts to discuss creature comforts. It isn't reasonable to assume that your crew members can stand on a hard concrete floor for hours on end during an operating session.

I therefore carpeted the step with strips of the same carpet tile that I used to carpet the entire basement, **16**. This carpeting comes in 22" x 23" tiles with mildly serrated front and rear edges that help maintain the alignment. Tiles are backed with nonskid foam padding that keeps them in place without resorting to messy carpet cement, and the backing is very easy on the feet. If the tiles get wet, as several of mine did on two separate occasions, they can be tossed in a dryer or carried outside to dry in the sun. If one is damaged by spilled paint or molten solder, it's easy to replace.

It's all about compromise

If you learn nothing more from this chapter and perhaps even this entire book, it's that model railroad design, and especially multi-deck model railroad design, is primarily about compromise. You can select an ideal set of Layout Design Elements (see my book *Realistic Model Railroad Building Blocks*) upon which to base your railroad, only to find that you can't include enough of them in a row to achieve the type of scenic or operational setting you had hoped for. Then comes compromise: less railroad, fewer LDEs, perhaps a smaller scale – or a multi-deck design.

Choose the latter and you're faced with still more compromises on layout height and scenic depth. In my view, as one who has built and is now operating a rather large multi-deck model railroad, the gains by far outweighed the more negative aspects of a multi-deck approach to layout design. Whether this will apply to you is for you to determine, but be aware that layout design is still about making compromises. At best, you get to choose which ones.

Speaking of compromises...

Choosing adequate lighting for your multi-deck layout is among the most important decisions you'll make. And, as chapter 7 shows, it's once again all about making reasonable compromises.

1

Bill Schneider

CHAPTER SIX

Layout lighting

The realism and enjoyment of a scene depends on how well it is lighted. Poor (dim, reddish, uneven) lighting makes it hard to see or operate a model railroad. Bill Schneider has ensured that this town scene on his New York, Ontario & Western HO layout is evenly and brightly illuminated from end to end with a twin-tube fluorescent fixture attached to the upper deck. He painted the underside of the top-deck structure sky blue to reflect light.

Lighting is perhaps the most important – yet poorly designed – support system on any model railroad, and that goes double for multi-deck layouts. If the layout is not designed from the outset to accommodate excellent lighting on the lower deck, fully half of the mainline runs of one's trains will be relegated to the shadows. Manually uncoupling cars will be difficult in the gloom. Those who think a string of white Christmas tree bulbs or lengths of rope light are adequate are deluding themselves. The resulting illumination is reddish, dim, and uneven. Those of us who are, or one day hope to be, 60 years of age and counting will have lost a significant percentage of our dim-light vision by then, and we'll need a lot of candlepower to compensate.

Safety first!

Please, if you're not well versed in wiring 120-volt AC circuits, hire an electrician to install your lighting and other AC power circuits. Like hiring a plumber to move a water heater that stands in the way of your proposed main line, hiring a professional electrician is money well spent. He or she can help ensure that you will meet all applicable local electrical code requirements, and the electrician may actually find ways to save you money while improving your lighting plan.

Types of lighting

In *Planning Scenery for Your Model Railroad* (Kalmbach Books), I noted that lighting deserves a chapter of its own, so here goes. As I noted there, it's a story that's still being written and is likely to change dramatically in the next few years. We are already seeing efforts to phase out the inefficient incandescent bulb that has been a mainstay of home lighting since Thomas Edison's day. Compact fluorescent bulbs are cited as the wave of the future, but they contain mercury and hence create disposal concerns.

What's next? Light-emitting diodes (LEDs), which will play a major role in home and layout lighting, are already coming onto the market, but at relatively high prices. We're already seeing them used extensively for vehicle lighting and even for railroad block signals and crossing flashers. Increasing use of LEDs in home-lighting applications will cause prices to drop, and they will offer energy-bill savings, less heat generation, and longer life (see www.ledlight.com).

For the short term, however, fluorescents, **1** and **2**, clearly offer numerous advantages – a linear light source along a linear right-of-way, more light per watt and hence less heat in the railroad room, longer life than incandescents, and choice of "flavors" (colors) of light, to name a few.

In fact, lighting designer Gerry Cornwell of Gerry Cornwell Lighting has given an eye-opening clinic in which he begins with a terse statement that fluorescents are the preferred method of layout lighting. He then convincingly demonstrates why that statement is true for most layouts (there is no universal solvent, of course)

Ceiling and under-cabinet fluorescent fixtures ensure that Jack Burgess' Yosemite Valley RR is not only brightly and evenly lighted but also that the entire room is illuminated with the same "color" of light. This makes it easier to get good color photos and avoid pockets of off-color lighting. The workbench and spray-painting areas also must have the same type of lighting.

Jack Burgess

The author experimented with mounting under-cabinet fluorescent fixtures a foot or so apart (above) vs. butting them as closely together as possible. The greater the distance between fixtures, the more apparent were shadows on the sky backdrop, especially on a narrow multi-deck shelf layout. Even when they were butted together, a slight but usually unnoticeable shadow remained.

and uses various colors and types of lighting to make his point. Any lingering doubts about the viability and, for the present, superiority of fluorescent lighting for most model railroad applications are left on the clinic-room floor.

I have used cool-white fluorescent tubes to illuminate two basement-size model railroads, and until LEDs are available, I see no alternative. Even the efficient fluorescents generate more heat than I would like. I now have 500 linear feet of railroad on two decks (three in places) to illuminate, and up to 18 people are needed to run the railroad during a four-hour session. This creates a lot of heat, even in the winter.

We retrofitted high-velocity central air-conditioning to our total-electric home in part to get rid of heat generated during operating sessions. No way could I deal with the heat generated by incandescents or quartz-halogen lighting, even if I ignored the spotty nature of any point source of light.

A brief review of physics

Perhaps no aspect of designing and building a model railroad generates a greater diversity of opinion than lighting. Like color preferences, it's highly subjective: What looks just spiffy to me may elicit a "What were you thinking?" reaction from others.

In a probably feeble attempt to introduce a science-based foundation for our discussions of lighting, let me offer a brief physics lesson: Lighting from a point source, such as a light bulb, drops off with the square of the distance. That is, if you move the object being illuminated (a model or your daily newspaper) from two feet to four feet away from a light bulb (double the distance), only one-fourth as much light will reach the object being illuminated.

Do the same thing with a linear source, such as a fluorescent tube, and you lose only half the original light intensity. If you have a "sheet" source of light, such as a ceiling filled from wall to wall with recessed fluorescent fixtures, there is virtually no drop in lighting intensity from the ceiling down to floor level.

In practical terms, this means a fluorescent tube bests a light bulb every time. At a given light-to-layout distance, you get about twice as much light from a tube as from a bulb, although there are many variables such as wattage that can change the equation.

I've heard some modelers complain that fluorescent lighting is too dim or even gray. That's usually because the fluorescent fixture was mounted too far from the scene being illuminated, or too few fixtures were used. The cure is to use twin-tube fixtures, which will usually provide the desired light intensity, or even to use multiple fixtures mounted close together, as in an office ceiling. Using a brighter fluorescent – in the 5,000K range (more on that in a moment) – should also help dispel such inaccurate views.

Point sources of light (bulbs) often create cones of light on a nearby backdrop and/or pools of light on the railroad, just like a streetlight does. This is especially true for multi-deck layouts, where the light source may be located close to the backdrop and the scenery.

Fluorescent tubes do much better here, but allowing even short gaps between fluorescent fixtures can create unwanted shadows on the backdrop at regular intervals. This is especially noticeable on lower decks, where fixtures tend to be closer to the railroad, **3**. The resulting shadows are easier to disguise on a serpentine mountain railroad, but they look like shadowy fence posts on the backdrop along a linear layout like mine.

Ken McCorry now uses a 1¼"-thick GE under-cabinet fixture that is linkable (up to ten fixtures), which greatly eases wiring chores, and comes in 12", 18", 24", and 36" lengths. The lighting is on multiple 120-volt circuits to keep the current load well below circuit-breaker ratings and to allow him to use only the lighting he needs when working on the railroad.

Ken uses 3"-thick L-shaped plywood brackets to hold up the upper deck, and he recommends cutting a notch for the under-cabinet fixtures where the brackets abut the valance/fascia, **4**.

Thin under-cabinet fixtures use T5 (⅝"-diameter) tubes in various short lengths. They are not as readily available as the more common T8 tubes (1") or T12 tubes (1½"), but are easier to get in standard colors such as cool white, another reason I opted to stick with CWs.

Standard under-cabinet fixtures, which I used wherever possible to illuminate the lower deck, **5**, are typically about 1¾" thick. Where I glued ¾" plywood splice plates to join the ends of eight-foot-long sheets of subroadbed, I used 1"-thick under-cabinet fixtures to compensate for the double thickness of the infrastructure.

Any type of under-cabinet fixture typically locates the center line of the fluorescent tube about 5" inboard of the valance. Since I locate the center line of the track 6" in from the aisle for elbow and derailment protection, this means that rolling stock or structures between the track and the aisle may be top- or even backlighted. This has not proven to be a problem, and in some ways it creates a shadow effect that adds visual depth to a scene, much like the New Hampshire barn in **6**.

Two photos: Brad Bower

These two photos of Ken McCorry's HO layout show how he mounted fluorescent strip lights behind the valance to ensure bright and even illumination on the lower decks. The deep upper decks are supported by one-piece L-shaped brackets cut from ¾" plywood.

4

Compact fluorescents

Allen McClelland used Sylvania 5,000-degree Kelvin screw-in compact-fluorescent (CF) "bulbs" to illuminate the second edition of his single-deck Virginian & Ohio, and they worked very well indeed, **7**. Layout width was relatively narrow outside of yards: I measured the typical distance from CF bulb to backdrop at 20", and the distance to the layout surface ranged from 28" to 38". Allen spaced the fixtures about 30" apart and used lower or higher wattage ratings as required, which proved to be more than sufficient to allow light from one CF to overlap that emanating from its next-door neighbors.

I initially assumed that as the lamp-to-layout distance decreased, as is typically the case for a multi-deck or narrow shelf-type layout, one would start to see cones of light on the backdrop, just as with incandescent bulbs. But recent experience with a double-deck extension to Perry Squier's HO layout showed that the CF lamps are long and large enough in diameter to act somewhat like linear sources, largely negating concerns about point sources. The lighting on his layout appears even despite the minimal 11" vertical clearance between decks, **8**.

More testing

Just as when picking a minimum radius, choosing a maximum grade, or establishing between-deck spacing for your layout, I recommend running a few lighting tests before making a commitment. Try several types of fluorescent tubes or CF lamps starting with cool whites and ranging through higher-temperature-rated tubes or lamps such as GE's Chroma 50 or Sylvania's C50 tubes to see what you prefer. The lighting is rated in Kelvins (as in 4000K), a temperature scale like Fahrenheit or Celsius, and the closer you get to the Sun's approximately 5600K, the whiter (closer to full daylight) the light, **9** and **10**.

The perceived "color" of lighting is highly subjective. The GE Chroma 50 tubes or their equivalent are used in areas where, for example, color-printing corrections are made, as they provide an almost pure-white light. When I tested them in my basement, however, they looked cold to my eye, and to my wife's, so I stuck with less expensive and easier to find (especially in the smaller sizes needed for under-cabinet fixtures) cool whites.

Similarly, warm-white tubes, intended to improve skin tones in office settings, are much too red for my tastes, causing pinkish smears on

Standard under-cabinet fixtures, which are about 1¾" thick, were mounted to the underside of the author's top deck. Where splice plates joined adjacent eight-foot sections of ¾" birch plywood subroadbed, he used 1"-thick fixtures.

the blue-sky backdrop. But they may look fine if you're modeling more arid climes like the American Southwest.

Also check the CRI – Color Rendering Index – to see whether that makes a difference to your eye. Santa Fe modeler Jared Harper recommends 5000-Kelvin tubes (like Chroma 50) with a CRI of at least 90. Such rating scales may be handy, but I recommend that you simply choose what is pleasing to your eye.

Upper-deck lighting

As we discussed in chapter 5, the height of the ceiling above the floor and the type of ceiling material can adversely affect choices for a layout lighting system. Concerns range from how to run wire between a circuit breaker and a string of light fixtures (and how many fixtures can be on a circuit) to how deep the fixtures are and how they can potentially affect headroom.

I screwed drywall to the main-floor joists, then mounted L-girders directly above the aisle fascias to hold the lighting valances for the railroad's upper deck. Fluorescent fixtures were then mounted at an angle within the L-shaped brackets, **11**. I placed the fixtures as closely together as I could to eliminate dim areas between fixtures, but the substantial distance between the ceiling and upper deck (it ranges from 18" to 30") minimized any between-fixture light fall-off.

Under-cabinet fixtures used to illuminate the lower deck had to be butted together to avoid casting shadows on the backdrop, **3**, owing to the reduced distance from fixture to railroad – 11" in downtown Frankfort to a maximum of 16" elsewhere.

I limited the number of fixtures on each of several separate lighting circuits to maintain a safe margin below each circuit breaker's rating. I also had an electrician install a separate branch panel to accommodate the railroad's electrical needs – mainly lighting but also wall outlets that provide power to run trains, operate switch motors, and so on. I can therefore shut off all circuits related to the railroad by pulling a breaker in the main entrance panel if need be.

When you can look across the layout to another aisle, you may also face the glare of fluorescent fixtures mounted close to the other aisle. Dan Holbrook solved this problem in a clever way without creating baffles that get in the way during model photography, **12**.

Ultraviolet damage concerns

Hang a photo or printed picture where sunlight can shine on it and you'll soon notice fading from the ultraviolet rays. The same thing happens to some extent with fluorescent lighting. I recall some early ground-foam scenery material that turned to brown powder from UV damage.

During the quarter-century that the Allegheny Midland resided in

Backlighting creates deep shadows on the visible side of this New Hampshire barn (above). A similar effect may happen to structures placed close to the aisle on lower decks owing to the top lighting from under-cabinet fixtures (right), but the shadows will help draw the eye into the more brightly illuminated parts of the scene, including the main line.

6

my basement, I could find not a trace of UV damage. Had I found clear evidence of UV-induced fading, I would have purchased the clear-plastic sleeves used by libraries and museums to ensure against such concerns. I opted to ignore such worries with the new NKP layout.

My judgment is reinforced by a comment recently made to me by lighting designer Gerry Cornwell: “I believe that for the majority of modelers, using UV filter tubes over fluorescent lamps is overkill. The amount of damage to objects from light sources, including daylight, is dependent upon these factors: sensitivity of the object, amount of light, amount of UV content of the light, and time of exposure.

“The most sensitive objects in museums are limited to 50,000 lux hours per year. That means 50 lux (5 foot-candles) for 1,000 hours or 50,000 lux for 1 hour. And these objects are typically much more sensitive to color changes than our models.

“A model railroad on public display with long open hours might benefit from UV filters, but most of our layouts are illuminated for a fraction of this time. And most paints and materials used in model making are quite robust.

“Finally, the vast majority of model railroads are poorly illuminated and would greatly benefit from higher quality lighting.”

Speaking of lighting quality, Ray Breyer points out that our eyes and ears may be very sensitive to the flickering and buzzing of poor-quality fluorescent fixtures. He therefore urges modelers to buy good-quality, name-brand lighting fixtures.

▲ Allen McClelland used a series of compact fluorescent lamps to light the second version of his popular Virginian & Ohio RR. The result was bright, even illumination and low power consumption (and hence low heat output).

▼ Perry Squier also used compact fluorescents to light the lower deck of a recent extension to his Pittsburg, Shawmut & Northern layout. Despite their close proximity to the lower deck, the lighting is relatively even.

Christmas-tree bulbs, rope lights

Don't even think of using Christmas-tree bulbs or rope lights to illuminate a lower deck. The former create dim red pools of light; the latter barely enough light for you to find them in the dark. A friend said he was going to overcome the admittedly dim lighting of white rope lights by running three parallel strings of them to illuminate his lower deck. He now has them for sale, cheap.

Moreover, Jack Burgess points out, the intensity and color of a layout's lighting should be identical on all decks, **13** and **14**. Using fluorescents in the ceiling and incandescents to light the lower deck is bound to create visual disharmony.

When I recommended against using rope lights during a presentation at a National Model Railroad Association convention in Madison, Wis., a modeler came up to me afterward and said I was right about white rope lights

9

Tommy Holt

but was overlooking the potential of blue rope lights to simulate a moonlit night.

I went home, bought several strings of blue rope lights, ran some tests, and found they do a very nice job of simulating a moonlit night. I am still experimenting with a continuous single string mounted behind the valance, but initial tests are very promising.

I know from decades of personal experience and visiting other model railroads that it's all too easy to get caught up in a funky scenic quest, however. My model railroad wasn't built to look romantic on a moonlit night, but rather to re-create the operating patterns and practices of its prototype in the steam-to-diesel transition era. As nice as the blue light looks, with structure interior lights and gleaming signals piercing the dark, busy operators may find such dim lighting more of a distraction than it's worth.

▲▼ Professional photographer Tommy Holt uses standard photo lights to achieve realistic lighting when photographing scenes, such as this portrait of the *California Zephyr* at Stockton, Calif. (above), on his HO layout. General layout lighting is provided by Chroma 50 fluorescent lamps (below).

10

Tommy Holt

▲ **Upper-deck lighting on the author's layout is provided by fluorescent strip lights attached at an angle to L-girders, which are bolted to the ceiling. In wide areas, such as yards, twin-tube fixtures are employed.**

Two photos: Dan Holbrook

Sky-blue removable panels normally shield the fluorescent tubes in Dan Holbrook's HO layout (top), but they can be removed to replace the tubes or for better access for model photography (above). **12**

For the theater majors

Let me close by once again stressing the subjective nature of lighting. Those of you who have training in theatrical lighting will know all sorts of ways to bring mood and drama to a scene by using various types of lighting, most often incandescent in nature. Your knowledge of what looks good to an audience and pleases the cast will guide you toward all sorts of impressive lighting feats on your model railroad.

My goal is somewhat different. I want to ensure that the railroad is brightly lit to the point that it's easy for crew members to read the numbers on freight and passenger cars as they're being set out or picked up. I want plenty of light in the narrow opening between boxcars so crews can accurately insert a pick to uncouple cars. I want enough lighting in the aisles to ensure that we can all read the print on the employee timetable, train orders, and waybills.

When it comes to realism, as in model photography, **9**, likely as not I will often shut off the layout lighting system and use tungsten lighting to create the look of a bright sun with strong, deep shadows. For operating sessions, realistic or dramatic lighting is not a major objective.

If you're building a layout primarily to entertain casual visitors with spectacular scenic and lighting effects, you won't like fluorescents. They're soft and thus fail to create realistic shadows. They're flat and hence do not create different moods in various scenes. They're hard and expensive to dim and seldom can be dimmed to zero without "strobing."

Those of us who use our modeling skills, such as they are, to create an environment in which we can re-create the operating patterns of the full-size railroads will therefore have strongly divergent goals from those who derive their main enjoyment from building models and placing them in realistic settings.

Both approaches to our broad-shouldered hobby are well worth the time, effort, and expense put into them. But for the specific purposes of this book – the design and construction of

Jack Burgess

▲ As is clearly evident in this photo of Jack Burgess' Yosemite Valley (above), it is very important to use exactly the same type, color, and intensity of lighting on all decks.

▼ Stephen Priest, editor of the NMRA's *Scale Rails* Magazine, and his wife, Cinthia (also a rail magazine and book editor), enjoy operating their HO Santa Fe layout, thanks in no small measure to the bright fluorescent lighting mounted in the ceiling and below the upper deck.

Stephen Priest

multi-deck railroads – the pendulum has swung in the direction of even, bright lighting over the entire layout. That's because the most common purpose of a second deck is to extend the mainline run, and that is usually a direct result of a desire to build more operating potential into a model railroad.

What else should we keep in mind as we contemplate whether and how to build a multi-deck model railroad? We'll discuss that in chapter 7.

1

Bill Darnaby

CHAPTER SEVEN

Don't let the top infringe on the bottom

Design the benchwork so that, to the greatest extent possible, low-hanging obstacles such as heating ducts and plumbing do not restrict headroom and so that support posts are buried in the scenery. Bill Darnaby minimized the inconvenience of ductwork above the aisle at East Yard on his Maumee Route by locating the low end of the railroad here.

When you begin building the infrastructure for your first multi-deck layout, you'll find that what works well on the lower deck doesn't work well at all on the upper deck. Moreover, designing and building a multi-deck model railroad is a bit like juggling a handful of billiard balls. It's as good an example of multi-tasking as you can find in our hobby. Focus on this without considering that and one of them will bite you every time. And even if you're really good at keeping all of the competing and complementary facets of model railroad design and construction in mind all at once, you may fail to do them in the right order.

The thickness of things

We'll get down to the basics of actually building a multi-deck railroad in the next chapter. For now, however, let's consider a few things that will affect, or be affected by, the addition of one or more extra decks.

On a single-deck model railroad, the only thing you have to contend with above the railroad is a low-hanging pipe, air-conditioning duct, **1**, or even the ceiling itself. On a multi-deck layout, it's like granting "air rights" above an existing structure: You're deliberately building what for all practical purposes is a second model railroad a foot or two above the lower one.

It follows that any device hung below the upper-deck subroadbed should be no thicker than the thinnest required piece of hardware that will be mounted there – usually the lighting fixtures.

Don't miss the point I tried to drive home in the introductory paragraph of chapter 6: Lower-deck lighting is probably the most poorly done aspect of multi-deck model railroads that I have seen. So it's important to start with the lighting plan and then design all other aspects of the layout's cross-section around that requirement.

Mounting, say, a Tortoise switch motor in its normal position is therefore problematic on the upper deck, as it sticks down several inches, **2**. But Circuitron offers a bracket that allows the motor to be turned 90 degrees, thus reducing its height requirements. A Tortoise can also be mounted remotely using this bracket.

Semaphore signal or crossing-gate actuating mechanisms create similar concerns, as they may project well down into the lower deck's viewing area. Fortunately, there is an increasing number of products that will operate a three-position block or train-order signal without taking much vertical space.

It pays to acquire a sample of each type of mechanism you're considering to see how much vertical space it requires, how easy it is to install and maintain, whether it can be operated remotely or from multiple locations (if

▲ Layout planners have to consider the vertical height of everything from light fixtures and trestles to signal mechanisms and switch motors (Lemaco on left, Tortoise on right) when designing the upper decks.

► Several companies make low-profile switch motors, such as this Switch Tender from Micro-Mark, or mounting brackets that allow motors to be mounted on their sides.

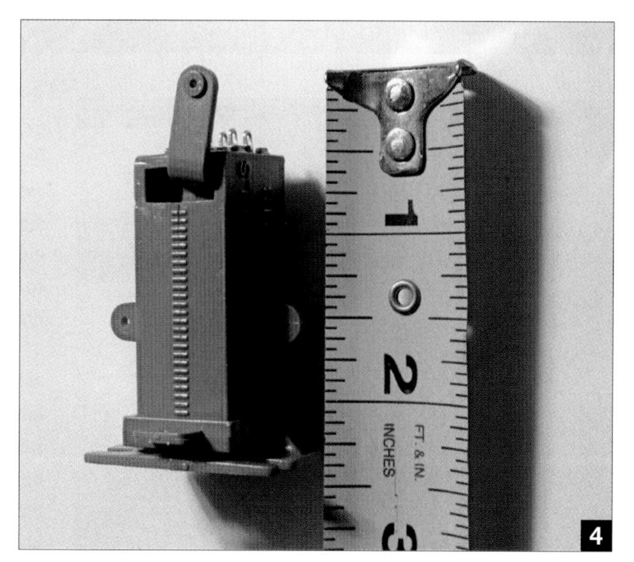

► New Rail Models makes the relatively compact (2⅜" tall) Blue Point manual switch-point controller. It's manually operated, which avoids the need for wiring, but routing the actuating cable around under-cabinet lighting fixtures may prove challenging on the underside of the upper deck.

Black-painted ⅛" hardboard skirting hides the peninsula's stud-wall backbone and the 2 x 2 legs that support the wider areas of the author's layout. This opening in the skirting provides ready access to lower-deck wiring and storage in plastic tubs.

▲ On one side of the central peninsula, the author's rail magazine library is stored in racks obtained at an office supply store. They're equipped with casters so they can be rolled out of the way during construction.

◄ Everything that is stored under the author's layout is in a homemade or commercial cabinet that can be moved to provide better access to the underside of the railroad.

desired), and how quietly it operates, **3**. I also recommend considering devices that operate switch points manually, such as the Blue Point actuator from New Rail Models, **4** (www.newrailmodels.com). What may seem like money wasted on samples now may pay huge dividends later on if your tests prevent making a significant design or acquisition error.

Signal sight lines

Giving road crews a clear line of sight in advance of wayside signals is a problem on any model railroad, but having decks above and below ideal viewing heights compounds the problem. Solutions range from relocating signals or the structures or scenic features that hinder seeing them to mounting repeater signal lights on the valance or fascia.

As with any other problem, knowing about such concerns in advance is usually sufficient to avoid the most glaring mistakes.

Below-benchwork access

We'll discuss how to gain access to upper-deck wiring in chapter 8 by running the feeders behind the backdrop to bus wires under the bottom deck. It follows that you'll need easy access to the underside of the bottom deck.

In wide areas of the layout where the benchwork is supported by legs that are covered by skirting, I have left a few open areas between skirting panels where I can crawl under the benchwork, **5**.

In other places where the benchwork is cantilevered out from the wall, I have installed commercial storage racks to which I added bottom braces and casters, **6**. These racks are usually available at office-supply stores, and each cardboard-divided section will typically hold a year's worth of model railroad or railfan magazines. With the addition of casters, I can roll them out of the way when I need to work under that part of the benchwork.

I have also purchased caster-equipped plastic cabinets for storing paint, brushes, electrical hardware, and paperwork, **7**. They, too, are easy to roll out of harm's way when I need to get under the railroad.

Hidden staging yards

If you make it difficult to reach a length of track, problems will occur at that precise spot in direct proportion to the degree of pain inflicted upon your person as you try to deal with them. Keep this in mind as you contemplate whether to build a multi-deck model railroad, as the likelihood of having to reach over and in to work on an upper deck, or the need to locate a staging yard (or helix) in a difficult-to-reach area, is considerably greater than with a single-deck railroad.

I had one small and three large hidden stub-ended staging yards on the Allegheny Midland, and I have three of them on the new Nickel Plate layout. The west-end (Fourth Subdivision) staging yard is in an open area at one end of the top deck, but it's 68½" above the floor – and hence close to the ceiling, **8**. The Peoria Division staging yard is located below the lower deck but is only six tracks wide, **9**, so access is not an issue.

The east-end staging yard comprises 17 tracks – the front 12 for the busy Sandusky Division and the rear five for the Toledo Division. It's located below the middle deck and hence is more difficult to access when things go wrong. The good news is that I can reach into this yard from the main aisle along the front and from the Charleston yardmaster's alcove along the back, **10**.

I installed rerailer track sections between every other section of flextrack to catch derailed wheels before things get out of hand.

At the last minute, I redesigned the east-end staging yard so that all 11 Sandusky Division turnouts are located along the main aisle and the four Toledo Division turnouts are along the aisle edge of the yardmaster's alcove, **11**. This allows the east-end crew to flip the Peco switches using a finger, at least until I install Tortoise switch motors. I have already installed switch motors for the five-track Toledo Division staging yard, as it is located out of reach of the main aisle.

8

Trains stored in the 68½"-tall, 12-track west-end (Fourth Subdivision) hidden staging yard on the author's layout are not normally visible but can be viewed and reached by standing on a step stool. Removing the valance and/or fascia makes it easier to handle maintenance chores.

9

The six-track Peoria Division staging yard, which feeds Frankfort yard, is tucked under Linden, Ind. All staging tracks are normally dead and activated by a push button on a panel near the entrance ladder.

10

The east-end staging yard accommodates trains to and from the 12-track Sandusky Division and five-track Toledo Division. It's accessible on both sides – from a main aisle and from the Charleston yardmaster's alcove.

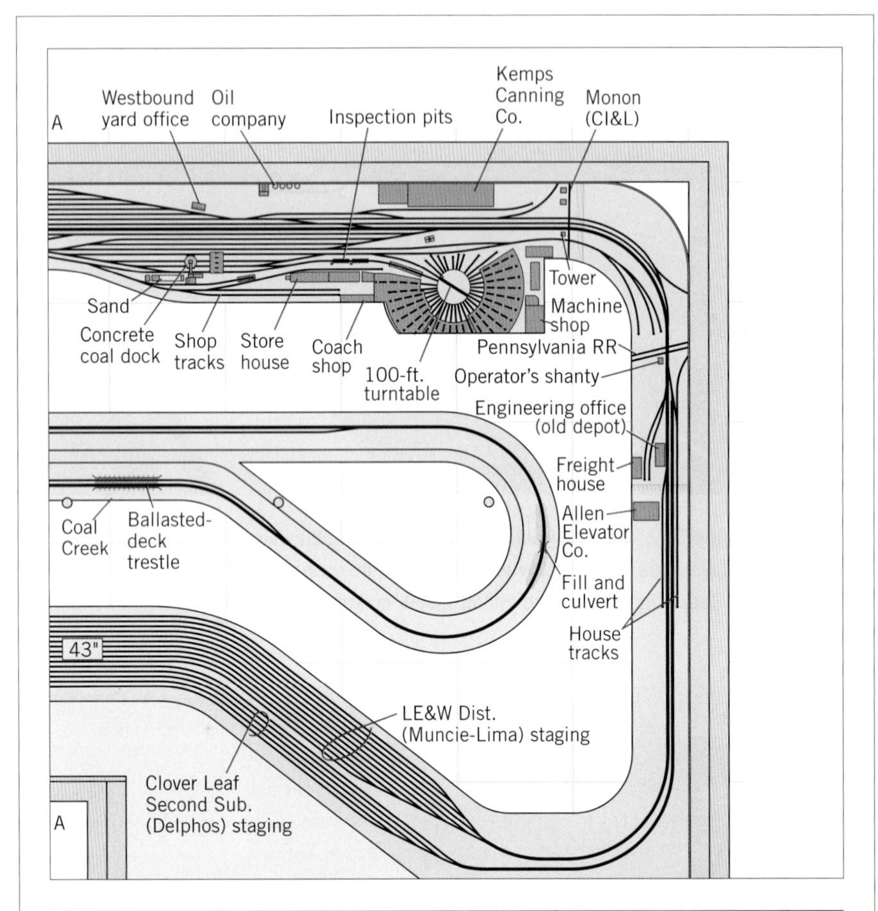

The east-end staging yard design (top) was revised during construction to locate the ladder turnouts next to the aisle to make maintenance or even replacement easier (above).

11

I prefer that all engine-storage tracks, including all staging-yard tracks, remain normally dead until needed. This prevents DCC decoders from needlessly receiving power, train crews do not need to remember to push function button 8 to turn DCC sound systems on or off, and it provides a small measure of added protection in the event of an over-voltage in the bus wires.

I do not trust busy crews to remember to turn toggle switches on and off, so I have installed push buttons on the three staging yard panels, **12** – the sole exception to my no-control-panel edict. A train crew must therefore stand attentively at the yard throat and push down on a button to power a track as his or her train enters or leaves a staging yard.

This, in turn, means that no staging track is normally powered, and current-detection devices cannot be used to show track occupancy. I therefore use two IRDOT infrared detectors (available in North America from Micro-Mark) or similar devices per track. One is located at the fouling point of each staging track's entrance turnout and another at the mid point of each track. These devices shine an infrared beam up to a mirror mounted on a signal-bridge-like L-bracket, **13**, which is reflected back to a detector; if nothing blocks the beam, the track is clear.

As I stage the railroad between sessions, I position each train just shy of the fouling point detector. An occupied track is denoted by a red light at the mid point of that track's schematic on the control panel. The second fouling-point LED should come on the moment the throttle is opened; if not, the crew knows to check the position of the reverse switch.

Trains entering a staging track first illuminate the fouling-point LED and then the mid-point LED. When the fouling-point LED goes out, the train should be stopped.

Office jobs

The motivation behind the construction of many multi-deck model railroads, including my own, was to get a longer mainline run while spacing a reasonable number of towns (and hence passing tracks) as far apart as possible for timetable and train-order operations. This implies that one will need a dispatcher and at least one operator to re-create the arcane but oh-so-fascinating practice of dictating and copying train orders in real time.

So where will they sit? This problem is exacerbated somewhat with a multi-deck plan, as there are likely to be more crew members in a given aisle at any one time because of the greater number of places to work and sidings for trains to meet or pass. Put another way, more railroad usually means more action. Having a desk for an operator or a dispatcher projecting out into a busy aisle is therefore even more problematic.

I wanted to have a dispatcher and two agent-operators, one for each main aisle. Realistic operation is increasingly assumed to mean that we're more realistically "modeling jobs" – not only those of engineers, yardmasters, and dispatchers, but also conductors, operators, station agents, towermen, engine hostlers, and even yard clerks and car inspectors.

Like his or her professional counterpart, the agent-operator not only copies train orders and messages but also checks with local industries to determine whether they need additional empty cars for loading. If so (as determined by comparing the desired quantity listed on a job aid or loaded-car waybills left in a bill box to the number of empties already on a shipper's siding), he or she prepares an empty car order and delivers it to the yardmaster. The yardmaster then strives to locate a suitable number of the desired type and grade of empties (say, boxcars suitable for grain loading) and sends them out on today's local.

It follows that the agent-operator needs a good-size writing surface on which to work, **14**. I located one in a wide spot in one main aisle near the Charleston yard office. I initially tucked the other operator's desktop under the lower-deck benchwork near the Frankfort yard office, but it and the operator's chair got in everyone's way. I have since relocated it to the far end of the second main aisle (see photo 9 in chapter 3), a much better location from both traffic and noise standpoints.

The dispatcher resides in the "blob" at the end of the central peninsula, a surprisingly spacious area inside the 42"-radius (the NKP's minimum) turn-back curve. He has to duck under the lower deck to enter his office (see page 27), but the floor is carpeted, and the main line has climbed to a 50" elevation (42" benchwork clearance) by this point.

Some modelers working locals prefer to work from switch lists. Dan Holbrook has therefore built

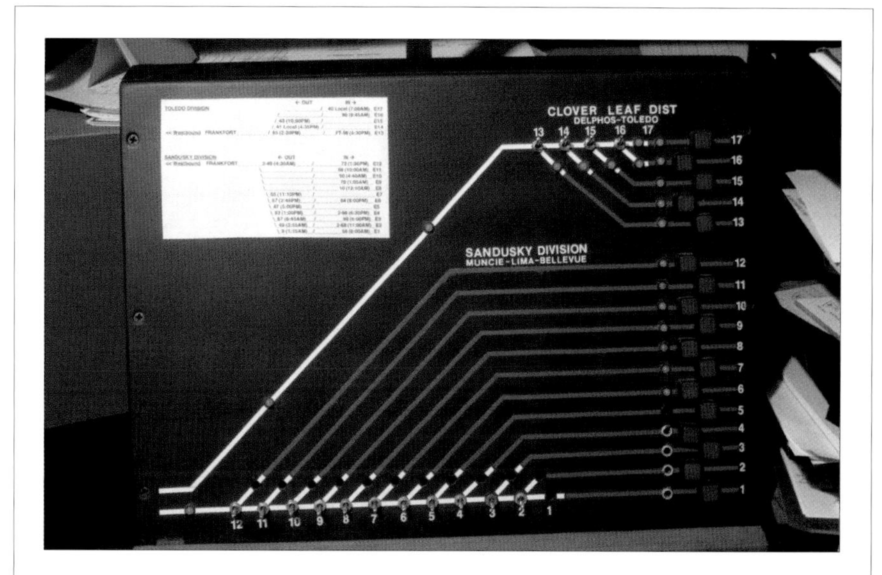

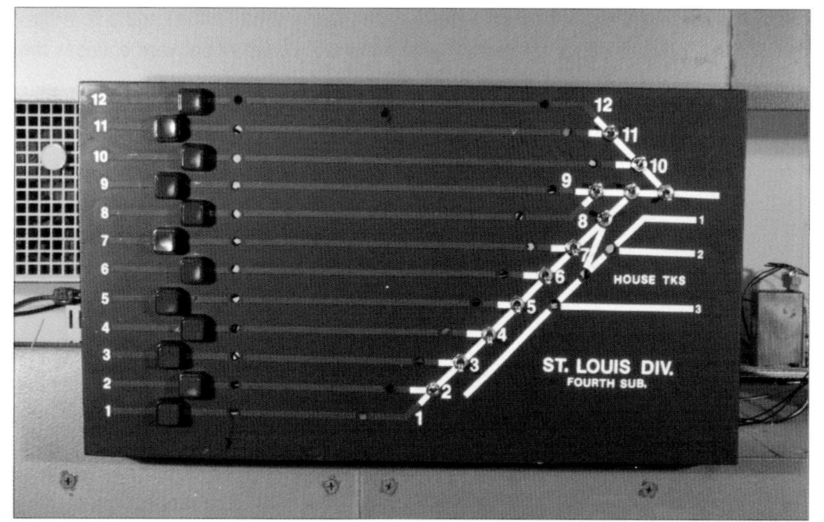

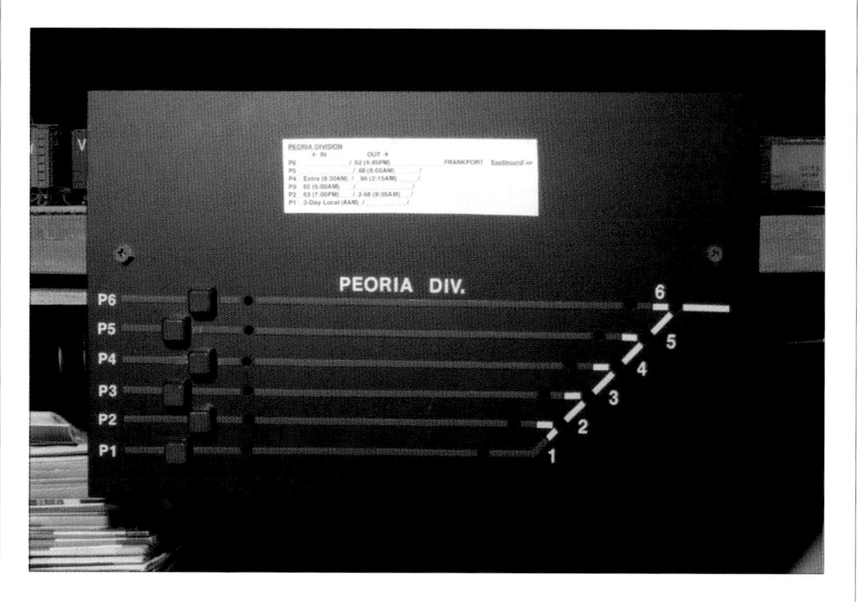

The under-construction control panels for the east-end (top), west-end (center), and Peoria Division (bottom) have SPDT toggle switches to control switch motors and SPST momentary push buttons to power each track. The LEDs show track occupancy.

12

Two IRDOT infrared detectors (above) per staging track illuminate LEDs on staging panels to show whether tracks are occupied and the fouling point of each turnout is clear. A small mirror on an L-bracket above each fouling point (below) and a 6" wide by 36" long "plastic" mirror located at the approximate center of the yard bounces the IR beam back to a detector if it's not blocked by a car or engine.

13

small versions of the corner desk of a caboose under the benchwork at several locations around his large Burlington Northern HO layout, **15**. A conductor can sit at one of these desks as he or she converts a stack of waybills into a switch list prior to heading out on the road. I'm now looking for a couple of places where I can tuck a small conductor's desk below the benchwork.

Those who roll their eyes at the idea of so much paperwork should pause to reflect on the realism this adds. Besides, what's the hurry?

Somewhat to my surprise, I find that the dispatcher and agent-operator positions, which handle most of the paperwork outside of waybills, are among the first jobs to be filled as I assign crews before each operating session. Early concerns about "no one will want that job" have therefore been quickly laid to rest. The idea that operation is primarily about running trains or switching yards has been put into proper perspective now that other interesting jobs are available.

What lies ahead?

Several years ago, I said that locomotive sound systems would become the norm – "No sound, no railroad!" is how I put it, which raised more than a few eyebrows – and that is clearly the case today. And it's not just the chuff-chuff of steam engines that adds to the fun; we now use whistle signals to call attention to a following section or send out or call in a flagman. In short, sound adds to the play value and realism at the same time.

A friend of mine, Jim Ferenc, recently dropped by to chat about progress on my Nickel Plate Road layout. I told him about how well the agent-operator positions were working out, noting that we no longer need telephones scattered about the railroad room, as train crews relay reports about passing each station to the proper operator, who then relays an OS (on-sheet) to the dispatcher: "Number 42 by Cayuga at 9:21."

Ideally, the agent-operator would note the passage of each train at the stations assigned to him or her, just as on the full-size railroad. But it's impossible to see every station from each operator's chair, I explained, and having multiple decks exacerbates the problem. Jim noted that he had rigged up some small TV cameras inside depots on his railroad, and the images they record are displayed on flat-panel screens, which could be disguised as a bay window, at each operator's desk. Fascinating! (See photo 13 in chapter 10.)

I've never been entirely comfortable pretending something is happening while my train sits dead still. I therefore installed Miller Models water

Bill Darnaby

▲ **One of the two agent-operator's desks on the Maumee Route is located below the roundhouse at Dacron, Ohio. The agent-operators copy and distribute train orders and messages from the dispatcher and work with the yardmasters to ensure that local shippers get empty cars for loading.**

► **Many conductors prefer to convert a stack of waybills into a switch list before heading out on a run, so Dan Holbrook constructed replicas of Burlington Northern caboose desks below the benchwork.**

Dan Holbrook

and coal loading sound modules by each coal dock and water tower. When the hostler spots an engine under the Frankfort coal dock, for example, the engine stays put until the water loading and then coal loading sound sequence is finished. Doing this with no sight or sound feedback is boring bordering on embarrassing.

Soon I expect Digital Command Control wizards to give us circuitry that knows by engine address the water and coal capacity of my steam fleet and shuts the engine down if time and power usage exceed an equivalent amount of coal and water – unless the engine has had those supplies replenished at a water tank or coal dock.

But I digress, so let's move on to a discussion of construction sequence and techniques in chapter 8.

1

Jack Burgess

CHAPTER EIGHT

Construction techniques

The challenge of upper-deck construction is to make the deck strong enough to support the railroad and accommodate a good lighting system without unduly infringing upon one's view of the lower deck. Jack Burgess did a masterful job on both counts while building his HO scale Yosemite Valley RR.

Let's consider basic wiring on the upper deck of your new railroad: The leads from the "north" and "south" rails no longer jut down into the dark recesses of the lower deck's benchwork but instead project down into the "sky" area above the lower deck. Installing the lower-deck's backdrop after roadbed and track are in place can be problematic. The upper-deck joists have double-duty to perform in that they have to be planned to accommodate the lower-deck lighting system. At times, it seems that every design decision on a multi-deck layout has the potential for unintended consequences. Clearly, we have some construction issues to wrestle into submission!

Upper-deck structure

The upper-deck structure obviously has some thickness, the vertical extent of which has to be minimized to avoid reducing the clearance between decks, restricting one's view of the lower deck, or forcing the upper deck to an even higher elevation, **1**. Conversely, the structure must be substantial enough to support not only the scenery, structures, and trains but also the lighting system for the lower deck – and probably the weight of someone leaning on it during construction or when rerailing a car.

You can readily see why the popular L-girder system of benchwork construction invented by former MR editor Linn Westcott is usually unsuitable for an upper deck: It's just too thick, **2**. Any system of benchwork construction from L-girder to the conventional grid is suitable for the bottom deck, but we need to be more innovative than that when it comes to the upper deck(s).

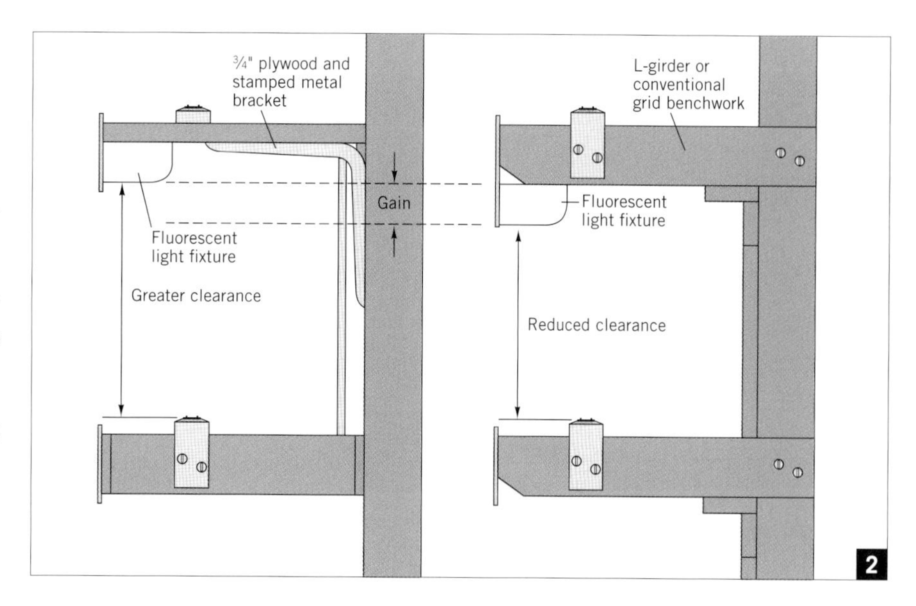

This sketch comparing a cross-section of L-girder or convention-grid benchwork shows how much thicker it is than the bracket-supported ¾"-plywood the author used for the upper deck of his HO layout.

Inexpensive stamped-metal L-brackets mounted on 16" centers support the top deck of the author's layout on either side of the central peninsula and along the long wall of the basement. Under-cabinet fixtures are attached to the underside of the plywood subroadbed outboard of these brackets.

Bill Darnaby employed a simple structure that uses 1 x 2 joists cantilevered out from the wall studs. These truncated joists then support the 2"-thick foam panels that he uses for subroadbed on the Maumee Route. I don't recall seeing any gussets to reinforce the stud/joist joint, but the joists are like airplane wing ribs – their close spacing (16") creates a rather strong and rigid structure.

Bill is happy with his foam-panel subroadbed (see the June 1994, March 1995, and April 1998 issues of *Model Railroader*), but I see a potential concern: It's thicker than, say, ¾" plywood, thus potentially reducing the clearance between decks by more than an inch if lighting fixtures are attached to the underside of the upper deck. And it's not easy to fasten a lower-deck lighting system to them.

I knew that I wanted to use under-cabinet fluorescent fixtures for lower-deck illumination, so I looked for a simple way to build a strong, thin upper deck. This led me to the use of stamped metal L-brackets, the type you can find at a lumberyard or home center for about a dollar each.

The largest commonly available bracket is about 12" x 14", which proved ideal for my typically 16"-wide shelf: It allowed a clear area beyond the end of the top arm of the bracket wide enough to accommodate the typical under-cabinet fixture, which measures 5" wide, **3** (and photo 5 in chapter 6).

Before I decided on this approach, however, I mocked up a section of the railroad's two-deck design. This showed me that, at typical elevations and deck separations, I could readily see the action on the lower deck, even though the top of a tall concrete grain elevator might be hidden behind the upper-deck fascia.

This "cropped-picture" effect was not a concern, however, as I had admired P-B-L's single-deck Sn3 display layout, which had a viewing area along the front that was perhaps 16" high. The tops of a grove of aspens, and perhaps the top of the wood coal dock, were truncated by the glossy-black valance, which somehow

Stephen Priest

Tall structures on a lower deck can be accommodated by shortening them or by extending them up behind the upper deck fascia/valance. Stephen Priest avoided the need to truncate the top of an elevator by recessing the upper deck on his HO Santa Fe layout.

enhanced rather than hurt the impact of the scene.

One way to avoid the cropped-top look is to recess the upper deck, **4** (also see photos 12 and 13 in chapter 3). I did this in two locations on the NKP: along the 63-foot length of Frankfort Yard and at Cayuga, Ind.

Cayuga presented an unusual engineering problem in that the sky backdrop bends around on either side of the double-track Chicago & Eastern Illinois main line and the wye connection to the NKP, **5**. Directly above this six-foot-wide opening is Charleston Yard. I built a supporting structure from 1 x 4s cut from ¾" plywood that resembles the spar and ribs of a wood airplane wing, and it has proven to be very strong with no noticeable deflection.

Along a short wall of the basement, the layout had to be 24" wide to accommodate the west-end staging yard. That led me to make the middle and lower decks (the latter for downtown Frankfort) the same width, and I knew the inexpensive stamped-metal shelf brackets wouldn't support the middle and upper decks.

I therefore purchased a number of 16"-long double-slotted Trak-Mount shelf brackets and the vertical channels that they snap into, **6** (www.johnsterling.com). The type I purchased has a slight downward slope from front to rear, which is not ideal for model railroad uses. I added shims to keep the subroadbed level, but I'd look for a brand that doesn't have this problem if I were buying more.

The middle deck in this area is on a constant grade, and the snap-in shelf brackets proved problematic in this regard. The slots into which each bracket snaps are spaced 1¼" apart, so it's impossible to position them at the exact elevation needed to maintain the desired grade. Once again, shims saved the day.

John Rogers ensured his layout shelves were rigid and level front-to-back by using steel carpenter's squares for shelf brackets, **7**. They aren't very expensive and can support relatively wide decks.

Jack Sibold used plastic brackets made by Rubbermaid (purchased at Home Depot), **8**, to support the back of shelves for his N scale layout. Dave Siegfried borrowed a design by Tony Steele for free-standing layout supports, **9**.

Planning wiring routes

It pays to plan well ahead when you're contemplating places to route feeder wires between rails on the upper deck and the main bus wires. Just running a pair of bus wires under the upper-deck subroadbed sounds easy enough, but that puts a lot of wiring right out where it may be seen by anyone, especially shorter folks, viewing the lower deck.

I therefore routed the feeder wires down through the ¾" plywood subroadbed, flush along the base of the subroadbed to the supporting wall, and then down behind the sky backdrop (⅛" hardboard) to reach the bus wires under the bottom deck. Take it from me: This is a lot easier to do before the backdrop is put up.

Having all of the bus wires under the lower-most benchwork makes it much easier to add feeders at a later date. Fishing later-installed feeders behind the backdrop may be a bit tricky, but connecting them to the buses will be a snap, especially if you use 3M's "suitcase" connectors (their formal name is insulation-displacement connectors, or IDCs) to join each feeder to its bus.

I used 3M connectors, **10**, on the Allegheny Midland's Coal Fork Extension with no problems. Neighbor Perry Squier used them on his basement-size HO layout without problems (he's a retired electrical contractor and recommended them to me), and I used them to wire the NKP. The brown connectors work fine to connect my 18-gauge feeders to 10-gauge bus wires, and I use red connectors to tie feeders to drop wires or switch motor wires to their buses.

Backdrops

Fitting the lower deck backdrop in place is much, much easier to do before any roadbed or track is in place. It's no chore to slide an eight-foot-long piece of hardboard or thick styrene in place along a straightaway, even with the roadbed and track in place, but it's almost impossible to do that on a turn-back curve at the end of a peninsula.

What makes this especially difficult is that you may need to notch the top edge of the backdrop every 16" or so to clear the built-in gussets in stamped-metal shelf brackets, **11**. This is no problem along a tangent, but the notch spacing may vary on curves, so I bent a scrap piece of fascia material around the curve and marked where I had to cut each bracket notch.

Unfortunately, I waited until I had the roadbed and track installed before trying to slide the eight-foot backdrop sections into position, which proved impossible to do on curves. I had to cut the hardboard strips in half and force them into place, then glue and spackle additional joints. Obviously, this would have been easier had I used a more flexible material such as styrene or linoleum.

The staged C&EI double-track main and "live" interchange track at Cayuga on the author's layout required a wide opening in the backdrop. To support overlying Charleston yard, he built an airplane-wing-like structure of spars and ribs from ¾" plywood.

I learned too late that I should have used a type of thin hardboard that has a glassy-smooth plastic coating on one side. It appears to resist expansion and contraction much better than standard hardboard. It's ideal for valances, as the white coating reflects light. I recommend it for valances, fascias, and backdrops with the plastic-coated side turned away from the aisle.

Note that the thickness of either the stamped shelf brackets or the double-slotted brackets will push the backdrop out toward the aisle by almost an inch, **12**. As long as you allow for this when planning your track and structure locations, it shouldn't present any problems.

So the recommended construction sequence is: lower-deck benchwork; main bus wires; upper-deck subroadbed; upper-deck backdrop; upper-deck roadbed and track; upper-deck feeders to the main buses; lower-

Commercial double-slotted shelf brackets, available in lengths up to two feet, were used to support an area along a short wall of the author's basement where the subroadbed expands to 24" wide on all three decks.

John Rogers cleverly used steel carpenter's squares to support the upper deck of his HO New England-based railroad. They're extremely rigid and can support wide decks.

Jack Sibold

Jack Sibold employed Rubbermaid metal shelf brackets to anchor the back of the subroadbed for his N scale railroad to the perimeter walls of the layout room.

Dave Siegfried

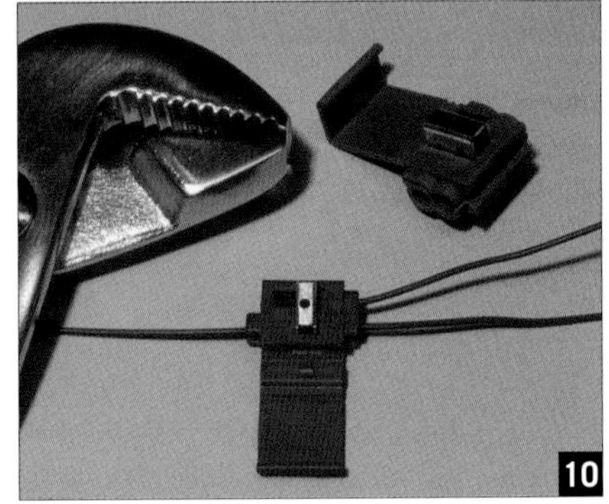

▲ The author used 3M's red insulation-displacement connectors ("suitcases") to connect drop wires to feeders and brown IDCs to connect feeders to bus wires.

◄ Dave Siegfried borrowed an idea from Tony Steele when he constructed freestanding supports for his layout, thus avoiding the need to attach the layout to the walls.

The lower deck's sky backdrop must either be cut short of the upper-deck supporting structure or, as the author did, notched to fit around the brackets. This is tricky on curves.

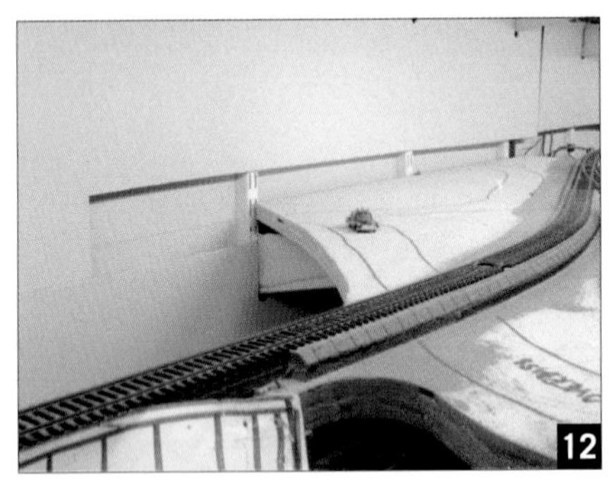

Double-slotted shelf tracks (visible here) or the gusset stamped into metal shelf brackets forces the backdrop material out from the wall about ¾", a loss that must be considered when locating scenery and structures.

deck backdrop; then subroadbed, roadbed, and track on the lower deck; and finally scenery and structures.

If you have defined your scenery during the planning phase to the point you know what goes where, I recommend painting the backdrop or installing photomurals before the roadbed, track, and scenery get in the way. This isn't difficult if you model a specific prototype, as you can accurately determine what each scene should look like well in advance of construction. In fact, you can photograph the actual scenes that will make up your backdrop, **13**, and have them ready for installation early in the construction process.

If you paint the sky portion of your backdrop, which is likely even if you use photos for the area below the tops of trees, I strongly recommend a light blue or gray color. A rich sky blue may appeal to your eye, but it won't reflect as much light as a hazy-blue or thinly clouded sky, and that can be critical on the lower deck. Other backdrop tips appeared in my book *Planning Scenery for Your Model Railroad* (Kalmbach Books).

Skirting

Unless your layout is mounted on brackets that extend out from the perimeter wall and therefore does not require legs for support, you'll need to think about how open you want the area under the layout to be. I wanted to use some of that area for storage, so I installed hardboard skirting in front of the 2 x 2 legs as shown in photo 5 on page 68.

I recessed the legs 9" back from the fascia to provide toe room for anyone standing up against the fascia. Attaching the hardboard skirting to the legs was then a simple matter.

We'll discuss painting the skirting and other trim in chapter 9.

Framing the railroad

No matter how successful we are with our modeling endeavors and how well we light the resulting scenes, we need to put our linear pictures in a frame, which is where valances and fascias come into play, as we'll discuss in chapter 9.

Mike Confalone

The effectiveness of a photo backdrop in expanding the apparent depth of a scene is apparent in these two foot-wide scenes (above and below) on Mike Confalone's HO railroad set in New England. The presumed scenic penalty of a narrow shelf layout is erased by carefully blending the 3-D and 2-D scenery.

Mike Confalone 13

CHAPTER NINE

Framing the picture – fascias and valances

A 9" hardboard valance reflects the cool-white fluorescent light onto the author's layout while keeping the glare out of crew members' eyes. The ceiling-mounted valance is directly above the upper-deck fascia except in wide areas (left) where it is aligned with the lower-deck fascia.

First, let's discuss a few additional terms. Just as paintings and photographs are framed to delineate their edges, the three-dimensional model railroad "pictures" we create benefit from similar delimiters. A fascia forms the lower part of the picture frame, just as a valance serves as the upper boundary of the picture frame and usually hides the lighting. On a multi-deck railroad, the lower deck's valance is also the upper-deck's fascia. Although not all modelers like valances – they may get in the way when shooting panoramic photos, for example – in my view, their merits far outweigh any shortcomings.

When less isn't more

A valance should be deep enough to hide lighting fixtures and prevent one's eyes from seeing the actual light source, be it fluorescent or incandescent. Its depth is therefore a function of ceiling height in relation to the height of the uppermost deck. On my railroad, that worked out to 9" as seen in **1**.

On a multi-deck layout like mine, there are at least two fascias and two valances, but the upper-deck's fascia is combined with the lower deck's valance, **2**. This dual-purpose strip of hardboard or perhaps lauan plywood has two jobs to do: hide the lower-deck lighting, "benchwork," and other hardware while framing one edge of both upper- and lower-deck scenes.

Some minimum depth for this intermediate valance/fascia is needed for it to do its job effectively and efficiently. The goal is therefore not to reduce the height of this combined valance/fascia to zero – it has to help frame both "pictures" – but rather to make it just deep enough to hide from direct view the joists or brackets that support the upper deck, any below-deck switch or signal motors and linkages, and the lighting fixtures for the lower deck. Conversely, the deeper it is, the more it impedes one's view of the lower-deck scene, **3**.

I tried several "thicknesses" and came to the conclusion that a 3" deep valance/fascia for the between-decks area is about right. I actually had to replace the center valance with one 3½" high when I decided to run a blue rope light just under the under-cabinet fluorescent fixtures used to illuminate the lower deck. The blue rope light is a source of "moonlight" for possible night operating sessions as described in chapter 6.

Hiding glare

Even rather dim lighting can bother one's eyes when looking directly into the light source. Light fixtures recessed into the ceiling prevent direct viewing of the light source, but I prefer closer and more direct lighting right above each deck. In any event, such fixtures cannot illuminate lower decks adequately.

2

The upper deck's fascia also serves as the lower deck's valance, and it acts as a picture frame to separate and accent scenes on both decks. The author followed David Barrow's example by painting valances and fascias a satin-finish "CTC machine" olive green.

3

Bruce Chubb

Bruce Chubb used wood paneling for his fascias and valances. He trimmed the upper deck fascias to minimize interference with viewing angles on the lower deck.

Since I put my light fixtures behind a valance, one has to work hard to see the fluorescent tubes. But at the end of the main peninsula where the lighting strips curve around, **4**, it is possible to see some ceiling-mounted tubes. If this proves troublesome over time, I'll diffuse the light with the same type of translucent plastic lenses that come with under-cabinet fixtures used on the lower deck.

I've already done that at Charleston yard, **5**, where the ceiling-mounted lighting along the main aisle shines directly into the eyes of the yardmaster and hostler as they work in the

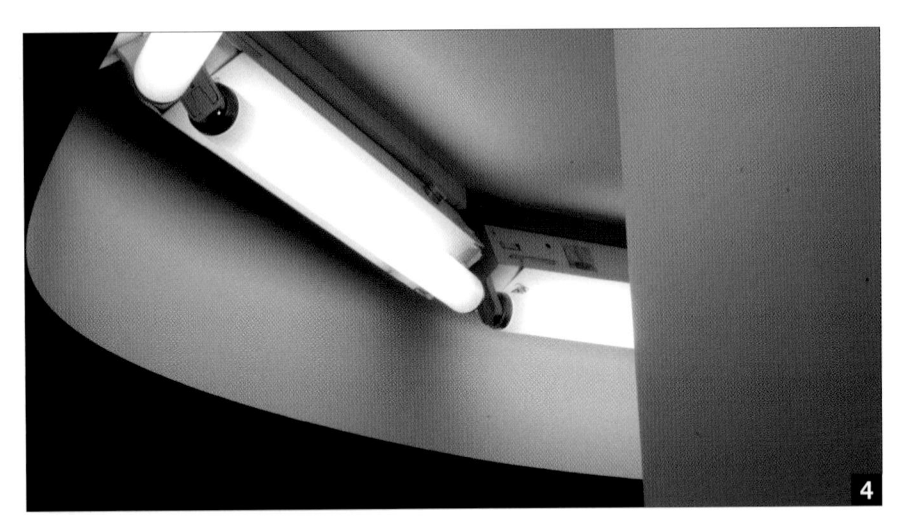

Fluorescent tubes are remarkably hard to bend, so the author used short fixtures on curves. The strip lights on the upper deck are visible from some viewing angles and therefore may be candidates for covering with plastic light-diffusing material.

The Charleston yardmaster works in an elevated alcove facing a main aisle, hence facing the strip lights along that aisle. Plastic diffuser panels block the glare.

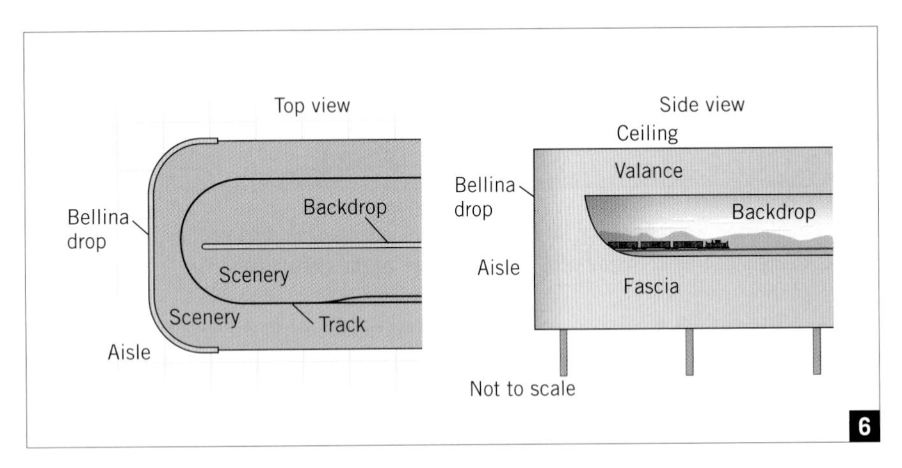

Jerry Bellina employed what we now call a "Bellina-drop" – a merged valance and fascia – to keep crews from standing at the end of peninsulas rather than following along with their trains. It also establishes a break between supposedly distant scenes.

elevated alcove. I simply swiped some translucent lamp covers from 48"-long under-cabinet fixtures that were mounted low enough for the lamp cover not to be needed and fastened them to the ceiling fixtures with good ol' duct tape.

The "Bellina-drop"

One of the pioneers of mushroom-style layout design was the late Jerry Bellina, who also designed the infrared Rail-Lynx command-control system. Jerry was determined to find a way to discourage train crews from standing at the invariably congested end of a peninsula as they watched their trains from a distance. Put more positively, he wanted to encourage crews to stick close to their trains.

Thus the Bellina-drop, **6**, a simple but effective idea: Extend the valance down to join the fascia at the ends of peninsulas, at scene breaks, and in other areas where you want to discourage folks from hanging around. Jerry's friend Craig Bisgeier is employing a Bellina-drop on his HO layout for the same reason.

One of the perils of multi-deck scenery design occurs when a vertically deep scene on an upper deck extends down into a lower-deck scene. On my layout, this occurred where a middle-deck bridge-over-river scene crowded into downtown Frankfort, Ind.

What first appeared to be a thorny scenic-design problem turned out to be rather easily resolved by employing the combined valance/fascia of a Bellina-drop. It dawned on me that a pair of towering grain elevators could hide the ends of a downward-projecting middle-deck fascia, and the fascia itself would nicely frame the sloping sides of the Little Vermilion River valley, **7**.

Although you're unlikely to encounter precisely this same dilemma, you will quite possibly have to accommodate a similar vertically deep scene on an upper deck. On a mountain railroad, a strategically placed tunnel on the lower deck could stand in for the grain elevators I employed. The interruption in the lower-deck lighting fixtures would be of no consequence here.

Nevertheless, some ingenuity will undoubtedly be required as you encounter depressions in upper-deck scenes. Planning their precise location before installing lighting fixtures may save a lot of head-scratching and hand-wringing later on.

Painting the trim

As long as we're discussing both skirting and fascias, let me share some thoughts about how they should be treated. Back in the 1970s when I built the Allegheny Midland, wood paneling was in vogue. That is no longer the case, so we have to determine how to finish our fascias, valances, and skirting.

In the 1995 *Model Railroad Planning*, David Barrow recommended painting the fascia and valance using Devoe's Upland Green, which is a close match to the pea green used on Union Switch & Signal CTC machines back in the steam era. I tried it on the newer Coal Fork Extension of the AM and quickly agreed with his choice. When I could no longer get this color, I had a local paint store match it in a latex satin finish (see chart), which doesn't show fingerprints as easily as a flat finish.

Fascia and Valance Color Mix

Sherwin-Williams Color Accents
latex satin Ultradeep Base

BAC	Colorant	Oz.	32	64	128
B1	Black	2	20	--	--
R2	Maroon	--	13	--	--
G2	New Green	--	16	--	--
W1	White	2	--	--	--
Y3	Deep Gold	2	32	--	--

(Matches Devoe Upland Green and is based on US&S CTC machine green; test before ordering large quantities!)

David also recommended painting the skirting black so the railroad seemed to be floating. That wasn't a concern on the AM, as I used wood paneling for everything, **8**, but I followed his advice on the NKP (see photo 5 in chapter 7). So did Tommy Holt when he built his Western Pacific layout, **9**.

Separating a vertically deep scene (top) on an upper deck from an urban scene on the bottom deck can be challenging. The author is taking advantage of tall grain elevators in downtown Frankfort, Ind., to frame the edges of the lower-deck scene and extending the fascia up to the valance to frame the bridge scene (above).

7

The clean look

You can get into some animated debates about what, if anything, to attach to the fascia or valance. Some prefer the ultra-clean look; after all, the combination of valances and fascias on a multi-deck layout is simply an extended picture frame. Clutter it up at your peril.

I like to mount photos of the prototype scene being modeled on the valance or fascia. This provides background information to casual visitors while documenting what you're doing to knowledgeable and appreciative modelers. If you choose to model something extraordinary, which I would question anyway – outstanding modeling is as much about plausibility as it is about highly developed modeling skills – documenting it with a nearby prototype photo may stave off a lot of inquiries.

Some modelers mount control panels for each town on the fascia. Bill and Mary Miller eliminated control panels as such and instead mounted a blueprint-like track schematic of each town at an angle on the fascia, **10**, so crews will know where to spot cars. To give the diagrams a railroady look, they created town names using Copperplate Gothic, a font available in most word-processing programs that was commonly used on railroad letterhead and business cards in the steam era.

Still others apply model airplane striping tape to create "full-size" track

The author used simulated wood paneling for the valance, fascia, and skirting for his Allegheny Midland RR. A more flowing, color-integrated look is more popular today.

Tommy Holt

Painting the skirting below the fascia black, as Tommy Holt did on his HO Western Pacific layout, gives the railroad a floating appearance. Using black curtains achieves a similar effect.

schematics along the fascia, taking care to locate the switch-point controls as close to the actual turnouts as possible to make it easier for operators to associate each toggle switch, push button, or knob with the device it controls.

Control panels

I am doing my very best to eliminate traditional control panels. They typically stick out into the aisle or cut into the scenicked portion of the railroad if they're mounted at an angle to make it easier to see what you're doing. If instead they're mounted close to vertical, then crews step back into and hence block the aisle as they try to get a better look at the panel. With few exceptions (interlocking tower "model panels" or CTC machines, for example), they're a purely model railroad device with no prototype basis. And they create a lot of extra wiring that has to be routed to a central location.

My upper-deck classification yard is switched by a yardmaster standing in an alcove out of the main traffic stream (see page 26 and photo 7 in chapter 5). But he could just as easily have been forced to work out in the main aisle. If so, where would I have located the control panel without having it block a main aisle and one's view of a significant part of Cayuga, Ind.?

One more point, and a key one at that when we recall the objectives I had for my new model railroad: modeling jobs. A control panel allows someone to throw a switch at a considerable distance. As long as you can see what you're doing, flipping a toggle here to move switch points way down there is considered fair game.

That may be realistic for an interlocked circuit on a CTC panel or an arrangement of armstrong levers in a tower, but for manually thrown switches, I'd rather the crew member walk down to the turnout and throw the switch right where a car or locomotive will soon be transiting. That's the exact place where things tend to go awry. This increases realism and safety while taking more time, and taking time to operate the railroad realistically is at the leading edge of prototype operation on today's better model railroads.

At interlocking plants that were controlled by armstrong levers or a small CTC machine, I am installing similar hardware for the station agent-operator to work, **11**. But at other locations, I am using either factory-installed (Micro Engineering and Peco) over-center springs to lock switch points in position, **12**, or I have installed micro slide switches. Moving the slide switch moves the points, and the SPDT contacts control frog polarity. I plan to cut off most of the slide-switch "knob" from each and mount a scale-size switch stand casting to it to hide the unrealistic knob, **13**.

An alternative is to use the operating ground throws made by Caboose Industries. One type has built-in contacts. If the switch points are spaced according to the NMRA's mechanical (not electrical) standards, I found you can use CI's smaller N scale ground throws to move the HO switch points, **14**.

Those who want to mount toggle switches or push-pull rods and knobs on the fascia, **15**, need to consider that the back of the upper-deck fascia (which is also the lower-deck valance) may be blocked by under-cabinet lighting fixtures. The solution may be to extend the fascia down another inch so the toggle switch or rod can be mounted below the fixture – and live with more restricted sight lines for the lower deck.

10

Paul J. Dolkos

A blueprint-like track schematic is conveniently mounted on the fascia at each town on Bill and Mary Miller's Colorado & Southern On3 layout to aid switch crews.

If you use push-pull rods, be aware that the actuating knob will project out into the aisle where it can easily be moved to an undesired alignment or bent by passing shoulders and elbows. The solution is to recess the knob in a "cup" mounted in the fascia, but again you have to plan ahead to be sure there is room behind the fascia for this indented housing.

The potential for bumping the knobs (or even toggle switches) is related to benchwork height. Those at hip or shoulder level are most prone to inadvertent contact.

Ready to move up in the world?

This concludes the "audio" portion of our discussion about the potential for and challenges of a multi-deck model railroad. I hope you now have enough insights as to what you may gain and the difficulties inherent in building a second or even third deck to make a well-considered decision as to whether

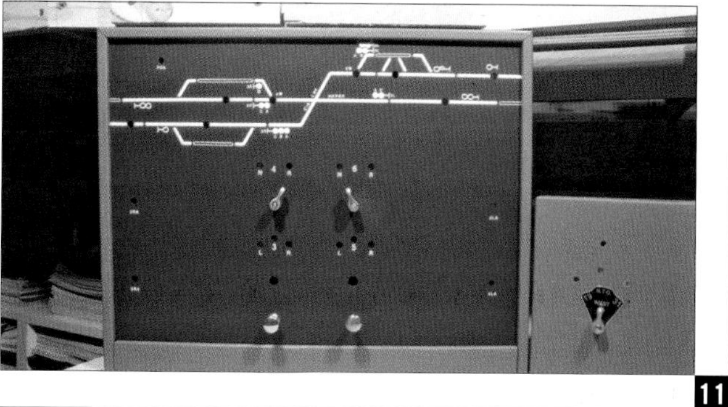

The author is building a same-size (12" x 17") copy of the simple interlocking panel that controlled the NKP-Monon crossing at Linden, Ind. The panel is hardboard with white striping-tape lines and Woodland Scenics dry-transfer Helvetica lettering; "T-O-O" butted together forms a signal symbol.

11

12

Micro-Engineering (left) and Peco turnouts feature over-center springs to secure the switch points for either route, thus avoiding the need for an electrical or mechanical mechanism to move them – and possibly a control panel to show which toggle switch or knob moves which set of points.

13

An SPDT micro slide switch can be used to move points mechanically while changing the polarity of an insulated switch frog. The author is investigating whether the "knob" can be trimmed off and an NJ International switch stand can be attached to its top surface with sufficient strength to withstand repeated push-pull movements. This is Perry Squier's HO layout.

14

Caboose Industries' relatively small N scale ground throws can be used to throw HO scale switch points if the point-to-stock-rail gap is set to the NMRA's mechanical (not the wider electrical) standard. This means the points have to be the same polarity as their adjacent stock rails, and the frog must be insulated.

15

Bill Darnaby

Bill Darnaby uses push-pull rods that move SPDT slide switches, which in turn move switch points and power the turnouts' frogs. The ball knobs are color-coded with white indicating the normal route side and red (mainline turnouts) or yellow (secondary-track turnouts) for the diverging route. This view of the Maumee shows Dacron (top deck) and Sciotovale on the right and Clay Center and Delphia at left. Note the knob at lower left coded white-outside/ red-inside, which means it should be pulled out (toward the white side) for the normal mainline route. Train-crew vigilance is required, as the projecting knobs can be bumped, but the switch-stand targets rotate as the switch points move.

this approach to model railroading best meets your needs.

Despite revolutionary work in this field by pioneers such as Jim Hediger, Jack Burgess, Jack Ozanich, Bill Darnaby, Joe Fugate, Jerry Bellina, Ken McCorry, and too many others to list here, the design and construction of a multi-deck model railroad is still regarded as somewhat of a pioneering adventure – hence this book.

It is therefore fitting that I devote the last chapter to photos of multi-deck model railroads. Their success should reassure you that you can indeed get there from here. I've asked these veteran modelers to shoot images that show their layouts "in context," as opposed to more realistic scenic or action photos. Their words and pictures will help you understand how they have applied the principles and tips covered in the first nine chapters.

1

Jack Burgess

CHAPTER TEN

Multi-deck layout sampler

This overview of the El Portal area near Yosemite National Park shows how Jack Burgess handled one single-deck section of his HO railroad. Note the fluorescent lighting in the drop ceiling, padded armrest, bead-board skirting, phone to the dispatcher, and fascia that blends in with California's golden grass.

Like almost any other aspect of model railroading worthy of special attention, the design of a multi-deck model railroad is part engineering and part art. Those who understand the mechanics of building a substantial yet efficient structure and can visualize two-dimensional plans in three dimensions have an edge. The rest of us will benefit from a study of their experiences. The following images of multi-deck model railroads suggest creative ways to deal with the multiplicity of mechanical and aesthetic challenges inherent in this increasingly popular but still challenging approach to model railroad design and construction. Enjoy!

Jack Burgess

The sky backdrop at Merced, on the left, extends up to hide upper-deck trains passing behind it on Jack Burgess's Yosemite Valley layout. The deep river scene on the top deck (distant middle) was located above a lower-deck area that has no tall scenic features.

Jack Burgess

The "thick" peninsula on the right contains a helix that connects the two main decks. The incline on the right goes up to the top deck. Note how Jack has blended the room colors from carpeting to painted backdrop to complement dried summertime grasses.

Bill Schneider

It's hard to imagine how Bill Schneider has managed to cram a lot of mainline railroading into a tiny bedroom yet achieve many wide-open scenes such as this one on his HO multi-deck New York, Ontario & Western HO layout. Photos 5 and 7 show additional views.

Two photos: Bill Schneider

If you're going to build a multi-deck layout, it helps to become proficient with photo-retouching software. The scene at left shows what a visitor would see, with a moving train to distract the eye from the ceiling. A little Photoshop magic shows how builder Bill Schneider hopes you'll remember the scene (right).

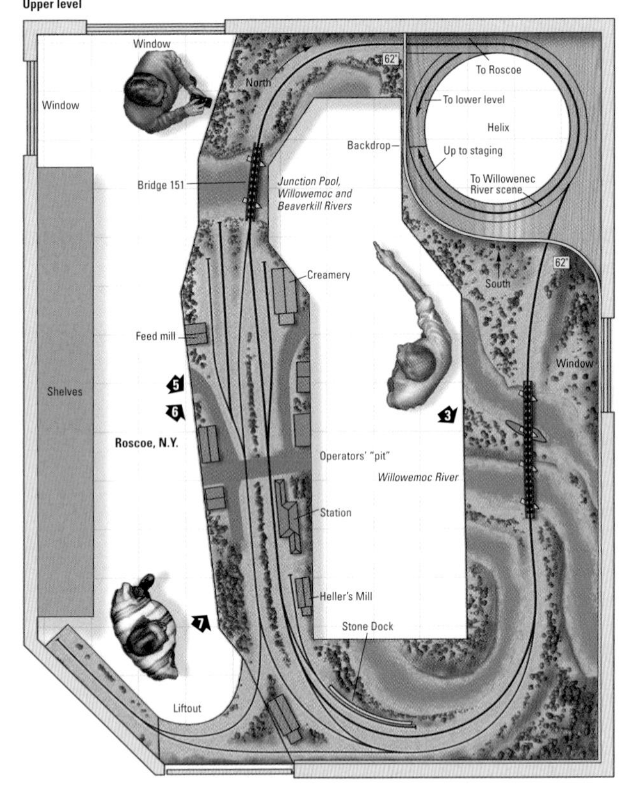

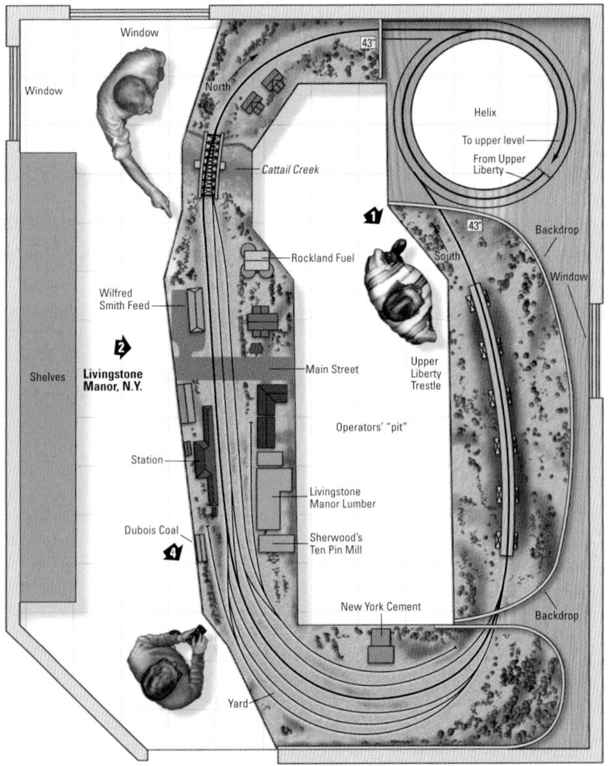

Illustrations by Theo Cobb

6

Compare these two photos to the track plan on page 90 – note the entry door at right – to see how much railroad Bill has managed to place in a very small bedroom without losing the expansive open bridge and small-town scenes that characterized the O&W. Despite the layout's small size, researching and modeling each town accurately provides plenty of challenging projects between operating sessions.

Two photos: Bill Schneider 7

8

John Longhurst

By recessing the middle deck on his CP Rail HO layout, John Longhurst has improved the viewing angle of the lower deck. Middle-deck viewing isn't unduly compromised, as its higher elevation allows almost straight-in viewing. The extended upper deck allows good lighting for the middle deck.

9

Bill Darnaby

To control the interlocking plant between the Maumee and two New York Central System lines at Edison, Ohio, Bill Darnaby built a small interlocking frame using Hump Yard Purveyance levers. It projects out slightly into the aisle. The clear boxes are waybill holders. An NYC diamond and tower actually existed at Edison; Bill simply superimposed the Maumee.

10

Ted Pamperin

Frankfort, Ind., on the author's newly operational HO NKP layout, is a busy place. A hostler manages motive power in the engine terminal, and separate yardmasters handle traffic in the westbound (far right) and eastbound (beyond the coal dock) yards. Humrick, Ill., on the upper deck (right) is recessed for improved access to the westbound yard.

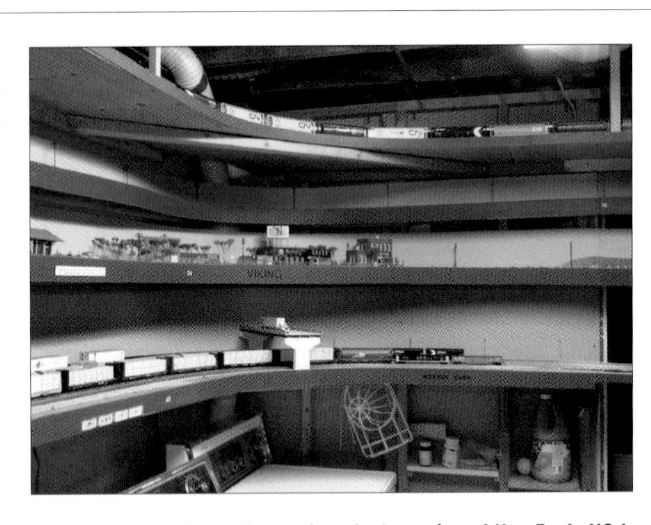

Two photos: John Longhurst

The two photos above show a four-deck section of Ken Epp's HO layout, with staging on the high top deck over the aisle. Ken kept the benchwork narrow above the washer and drier. He coped with the deep bridge scene at right by putting it on the bottom deck.

11

Bill Darnaby

Bill Darnaby

John Swanson

The top two photos show how very narrow scenes – essentially, only the railroad's right-of-way – on Bill Darnaby's Maumee nevertheless convey more than enough information to establish this as a Midwestern Corn Belt railroad. Bill used electrostatic grass along with other scenic products to achieve the natural-looking texture. He used Micro Engineering code 70 flextrack with handlaid turnouts. The photo at left on John Swanson's HO layout makes the same point: A narrow right-of-way shelf is sufficient to put the railroad into context.

12

Dan Holbrook

A comfortable and visually appealing environment goes a long way toward making construction and operating sessions more inviting. Dan Holbrook used dark fascias and valances to set off the railroad and complemented that with brown-trimmed, buff-colored bead-board paneling. Note the TV monitor in the "window" by the desk.

Tom Stolte

Seven-footer Tom Stolte built part of his multi-deck railroad high enough to clear the entry door at the back of this photo. The lower deck is at 54", the upper around 72". Access to the rest of the room is via the lift-out hatch visible to the right of the door. Tom models the Missouri Pacific in HO.

An S-curved backdrop on Perry Squier's Pittsburg, Shawmut & Northern provides mainline access to the helix between decks while hiding the dispatcher's office inside the helix. The knobs control SPDT slide switches that in turn move the switch points. Mr. Plaster brick kilns provide car loadings.

Three photos: Ann Lewis

These three photos of Mike McBride's under-constructon Chicago & North Western and Litchfield & Madison HO layout show how he created a flowing fascia by using spacers between it and the benchwork framing. As the photos show, he achieved equally flowing trackwork by using large-radius curves. "Portholes" in the fascia, cut with a fly-cutter in a drill press, provide staging access.

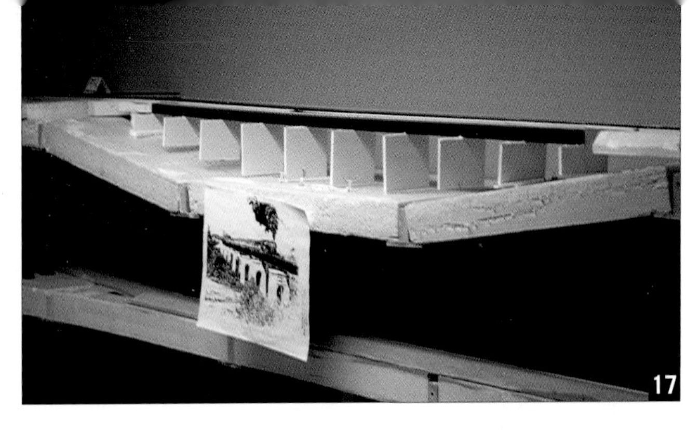

17

Paul Miklos is modeling the Baltimore & Ohio's Chillicothe, Ohio, line on a multi-deck N scale layout. He used 2" foam panels supported by T-shaped joists, which were lowered to accommodate the river in this bridge scene.

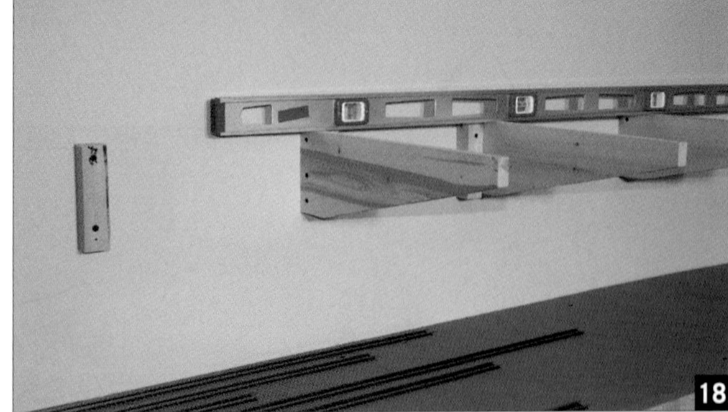

18

Bill Sornsin

Bill Sornsin is building a large multi-deck version of the Great Northern in HO scale. He erects the sky backdrop first, attaches short 2 x 2s, and finally adds the tapered 1 x 6 joists. "It's easier to achieve the desired level this way compared to screwing commercial brackets to the wall," Bill reports. The taper improves lower-deck viewing.

Four photos: Jim Hediger

It's appropriate to end a book on multi-deck layouts with photos of the first one ever built: Jim Hediger's Ohio Southern. The backbone of the central peninsula is the X-shaped supports visible at bottom left. Jim then attached L girders and set joists atop the girders. He built the benchwork from the bottom up and scenery from the top down to "keep the slop off a finished lower deck." Jim's pioneering application of John Armstrong's creative suggestion set in motion a trend that is now recognized as an effective way to double or triple the amount of railroad in the same footprint.

19